Your Body
The Missing Manual®

Your Body: The Missing Manual

BY MATTHEW MACDONALD

Published by O'Reilly Media, Inc., 1005 Gravenstein Highway North, Sebastopol, CA 95472.

O'Reilly books may be purchased for educational, business, or sales promotional use. Online editions are also available for most titles (*safari.oreilly.com*). For more information, contact our corporate/institutional sales department: 800.998.9938 or corporate@*oreilly.com*.

Editor: Peter McKie

Production Editor: Nellie McKesson

Copy Editor: Patricia Clewell

Indexer: Jan Wright

Cover Designer: Karen Montgomery

Interior Designer: Ron Bilodeau

Print History:

July 2009:   First Edition.

ISBN: 978-0-596-80174-8

[F]

Contents

The Missing Credits . . . . **vi**
Introduction **1**

Part 1: From the Outside In

Chapter 1
Skin: Your Outer Layer. **7**
    The Protective Wrapper 8
    Skin and the Sun. 13
    Going Deeper. 18
    The Oil Factory 21
    Your Temperature Control System 23
    Hair. 30

Chapter 2
Fat **37**
    The Purpose of Fat. 38
    Fat Up Close. 39
    Taking Stock of Your Body Fat 44
    Where Fat Lives. 51
    Understanding Your Body's Anti-Starvation System 55

Chapter 3
Muscles **65**
    Meet Your Muscles 66
    Exercise. 72
    A Grab Bag of Strength Exercises 80

Chapter 4

Bones **89**

The Skeletal System. 90

Living Bone 92

Joints 95

Your Spine 98

Chapter 5

The Doors of Perception **105**

Vision 107

Hearing. 115

Smell. 124

Taste 127

Part 2: The Machinery Inside

Chapter 6

Your Lungs **133**

The Respiratory System 134

Breathing. 138

The Path to Your Lungs 140

Your Voice 145

Pollution 149

Chapter 7

Your Heart **157**

The Pump Inside. 158

The Circulatory System 164

Heart Attacks 168

Cardio Exercise 172

Chapter 8

Your Digestive System **179**

In Your Mouth 180

Your Stomach. 186

Your Small Intestine. 190

Your Large Intestine. 195

Chapter 9

Your Immune System **205**

    Self-Defense. 206

    Bacteria 212

    Viruses. 220

    Cancer 226

    Flaws in the System 229

Part 3: Sex and Death

Chapter 10

Sex and Reproduction. **235**

    Male Sex Organs 236

    Female Sex Organs 243

    The Main Event. 252

Chapter 11

Your Final Exit: Aging and Death **257**

    Life Expectancy. 258

    Why We Die 261

    Aging 265

    The Anatomy of Death. 269

The Final Word **275**

Index **277**

The Missing Credits

About the Author

 Matthew MacDonald is a science and technology writer with well over a dozen books to his name. He's also the author of *Your Brain: The Missing Manual*, a quirky exploration into the odd and wondrous world of your squishy gray matter. In a dimly remembered past life, he studied English literature and theoretical physics.

About the Creative Team

Peter McKie (editor) is an editor at Missing Manuals. During the course of editing this book, he rejoined his gym, changed his eating habits, and started walking rather than riding (to the dismay of local cabbies, who have kids to put through college). He was graduated from Boston University's School of Journalism and lives in New York City, where he works and researches the history of old houses and, every once in a while, sneaks into abandoned buildings. Email: *pmckie@gmail.com*.

Nellie McKesson (production editor) lives in Brighton, Mass., where she spends her spare time (between working full time at O'Reilly and studying Graphic Design at MassArt) making t-shirts for friends (*http://mattsaunders-bynellie.etsy.com*). While working on this book, she took the plunge and went *no-poo* (it's probably not what you think: *http://www.instructables.com/id/How-to-Go-No-Poo/*). Email: *nellie@oreilly.com*.

Patricia Clewell (copy editor) lives in St. Louis, Missouri, where she works, makes mosaic art, and rides too many roller coasters with her 13-year-old son, Ben.

Jan Wright (indexer) lives in the mountains of New Mexico, and likes to ride an electric bike when she is not indexing, beading, or eating red chile. (*http://www.wrightinformation.com*; email: *jancw@wrightinformation.com*).

Susan Allison (technical reviewer) completed a PhD at the Garvan Institute of Medical Research in Sydney, Australia, investigating novel neural mechanisms of bone formation. She then undertook postdoctoral research in the field of neurogenesis at the Stem Cell Center, Lund University Hospital, Sweden. She is currently an editor of a medical journal in London.

Garrett Hurst (technical reviewer) is a third-year medical student at the Medical University of South Carolina. Raised in Columbia, South Carolina, he was graduated from Clemson University in 2006 with a BS degree in health science. He will graduate in 2011 with an MD degree and plans to pursue a career in the primary care field. Garrett has been involved in a variety of extracurricular service and social activities at MUSC, and in his free time he likes to play the piano, work on cars, and spend time with friends and family.

Acknowledgments

This is the part of the book that most readers skip on the road to the juicy stuff inside. If that describes you, thank you for buying the book, enjoy the trip, and please carry on without further delay.

If you're still here (Hi, mom!), you're probably one of the special few individuals who helped transform this book from unfocused inspiration into the odd little jewel that it is. Allow me to extend my thanks. For though I could have done it without you, the results would have been ugly. (Maybe not career-endingly ugly, but let's just say the book wouldn't feel out of place sitting between the professional-wrestler autobiographies and the Y2K remainder bin.)

The first in my long list of thanks goes to my dedicated reviewers, Susan Allison and Garrett Hurst, who offered helpful advice and worked industriously to catch my mistakes. Up next is my editor Peter McKie, who helped keep the schedule on track, and—on more than one occasion—climbed inside my head to help clarify a murky passage. Third, I extend a special thanks to the team of people who helped create the art for the book, including Joanna Lundeen, Belle Keller, Rob Romano, and all the talented artists at Dartmouth Publishing. As you can see, the end result is art that has the perfect mix of simplicity, insight, and accuracy. I also owe thanks to Nellie McKesson, who headed production and is responsible for the slick integration of picture and print, and to everyone else who worked to get this book formatted, indexed, and printed. You can meet many of them of the Missing Credits page.

Lastly, I thank my family—particularly my parents, who were in charge of the care and feeding of my body for the first decade or two of my life (and who, in conjunction with my wife's parents, are occasionally still responsible for it in the current day). I'm also eternally grateful for my wife Faria and my daughters Maya and Brenna, who worked tirelessly to devise ingenious ways to distract me when I should have been writing. My hope is that someday they'll read this book, and reclaim some of the time it took from them.

—Matthew MacDonald

The Missing Manual Series

Missing Manuals are witty, superbly written guides to computer products that don't come with printed manuals (which is just about all of them). Each book features a handcrafted index; cross-references to specific pages (not just chapters); and RepKover, a detached-spine binding that lets the book lie perfectly flat without the assistance of weights or cinder blocks.

Recent and upcoming titles include:

Access 2007: The Missing Manual by Matthew MacDonald

AppleScript: The Missing Manual by Adam Goldstein

AppleWorks 6: The Missing Manual by Jim Elferdink and David Reynolds

CSS: The Missing Manual, Second Edition by David Sawyer McFarland

Creating a Web Site: The Missing Manual by Matthew MacDonald

David Pogue's Digital Photography: The Missing Manual by David Pogue

Dreamweaver 8: The Missing Manual by David Sawyer McFarland

Dreamweaver CS3: The Missing Manual by David Sawyer McFarland

Dreamweaver CS4: The Missing Manual by David Sawyer McFarland

eBay: The Missing Manual by Nancy Conner

Excel 2003: The Missing Manual by Matthew MacDonald

Excel 2007: The Missing Manual by Matthew MacDonald

Facebook: The Missing Manual by E.A. Vander Veer

FileMaker Pro 9: The Missing Manual by Geoff Coffey and Susan Prosser

FileMaker Pro 10: The Missing Manual by Susan Prosser and Geoff Coffey

Flash 8: The Missing Manual by E.A. Vander Veer

Flash CS3: The Missing Manual by E.A. Vander Veer and Chris Grover

Flash CS4: The Missing Manual by Chris Grover with E.A. Vander Veer

FrontPage 2003: The Missing Manual by Jessica Mantaro

Google Apps: The Missing Manual by Nancy Conner

The Internet: The Missing Manual by David Pogue and J.D. Biersdorfer

iMovie 6 & iDVD: The Missing Manual by David Pogue

iMovie '08 & iDVD: The Missing Manual by David Pogue

iMovie '09 & iDVD: The Missing Manual by David Pogue and Aaron Miller

iPhone: The Missing Manual, Second Edition by David Pogue

iPhoto '08: The Missing Manual by David Pogue

iPhoto '09: The Missing Manual by David Pogue and J.D. Biersdorfer

iPod: The Missing Manual, Seventh Edition by J.D. Biersdorfer and David Pogue

JavaScript: The Missing Manual by David Sawyer McFarland

Living Green: The Missing Manual by Nancy Conner

Mac OS X: The Missing Manual, Tiger Edition by David Pogue

Mac OS X: The Missing Manual, Leopard Edition by David Pogue

Microsoft Project 2007: The Missing Manual by Bonnie Biafore

Netbooks: The Missing Manual by J.D. Biersdorfer

Office 2004 for Macintosh: The Missing Manual by Mark H. Walker and Franklin Tessler

Office 2007: The Missing Manual by Chris Grover, Matthew MacDonald, and E.A. Vander Veer

Office 2008 for Macintosh: The Missing Manual by Jim Elferdink

Palm Pre: The Missing Manual by Ed Baig

PCs: The Missing Manual by Andy Rathbone

Photoshop Elements 7: The Missing Manual by Barbara Brundage

Photoshop Elements 6 for Mac: The Missing Manual by Barbara Brundage

PowerPoint 2007: The Missing Manual by E.A. Vander Veer

QuickBase: The Missing Manual by Nancy Conner

QuickBooks 2009: The Missing Manual by Bonnie Biafore

QuickBooks 2010: The Missing Manual by Bonnie Biafore

Quicken 2008: The Missing Manual by Bonnie Biafore

Quicken 2009: The Missing Manual by Bonnie Biafore

Switching to the Mac: The Missing Manual, Tiger Edition by David Pogue and Adam Goldstein

Switching to the Mac: The Missing Manual, Leopard Edition by David Pogue

Wikipedia: The Missing Manual by John Broughton

Windows XP Home Edition: The Missing Manual, Second Edition by David Pogue

Windows XP Pro: The Missing Manual, Second Edition by David Pogue, Craig Zacker, and Linda Zacker

Windows Vista: The Missing Manual by David Pogue

Windows Vista for Starters: The Missing Manual by David Pogue

Word 2007: The Missing Manual by Chris Grover

Your Brain: The Missing Manual by Matthew MacDonald

Introduction

You know it intimately. Every morning, you climb out of bed with it and give it a good, hard look in the mirror. You scrub it down with soap, fill it up with calories, and occasionally walk it around the block for some fresh air. Most of the time, it's as comfortable as an old pair of jeans. But occasionally, particularly when something goes wrong, you realize just how much you depend on the nearly miraculous abilities of your human body.

Your body is a very odd place. It's at once mind-bogglingly complex, supremely adaptable, and hopelessly quirky. Often, it amazes. (For example, in the time you've spent reading this introduction, your body has created several *million* new cells to replace those that have just popped off.) Other times, it's somewhat less impressive. (Just try laughing while drinking a glass of milk.) But whether it's dazzling or disappointing, the human body ranks as the most remarkable machine you'll ever operate.

In this book, you'll learn how your body works, and how you can keep it in tip-top shape. You'll discover how to spot potential problems, and you'll learn to love—or at least live with—its charming and not-so-charming oddities. Along the way, you'll solve a few mysteries. You'll see why we use the same hardware to talk, eat, and kiss. You'll meet the millions of uninvited guests that share your body with you. You'll also learn why at least one study suggests that you'd be better off preparing dinner on a toilet seat rather than on an average kitchen cutting board (page 215). You'll face the ravages of aging, confront the challenges of dieting, explore the riddles of diseases like cancer, and learn what happens when you die. (Unfortunately, this book can't tell you much about what goes down after that.)

We think you'll enjoy the tour. After all, you'll be spending a lot of time in your body. Why not get better acquainted?

About This Book

This book provides a practical look at how to get the most out of your body. What makes this book different from the average self-help guide is the fact that it's grounded in modern-day biology—that means hard-core sciences like *biochemistry* and *genetics*.

Of course, there's no sense burying you in an avalanche of graduate-school theory. (If that's the experience you want, have a stiff drink, lie down, and hope the feeling goes away. If that doesn't work, call the nearest center of higher learning and maybe they can talk you out of it.) Instead, this book teaches you just enough biology for you to understand current research and pick up some practical body-boosting tips.

As you'll discover, a dash of science can explain quite a few of the stranger details of day-to-day life. For example, simple biology can explain why you get goose bumps when you're cold (page 31) and how—or if—antibacterial soaps really work (page 216). In short, the goal of this book is to give you the scientific insight you need to better your body. You won't, however, pick up enough science to clone dear, departed Uncle Stan.

About the Outline

Your journey into your body wends its way through 11 chapters:

- **Skin: Your Outer Layer (Chapter 1)** looks at your miraculous body wrap. You'll learn how dead skin cells defend your body and litter your house, what causes everything from a mole to a wrinkle, and how antiperspirants tame body odor.

- **Fat (Chapter 2)** sizes up one of the least welcome body ingredients: fat. You'll learn why you need it, where it hides, and how your body stores it. You'll also learn to distinguish healthy diets from the body abusers, and find out whether you're pre-programmed to pack on the pounds.

- **Muscles (Chapter 3)** considers your body's movers and shakers. You'll learn how muscles work, why you need to tear them up to make them grow, and how weight training can improve the health of your entire body (whether you're a man or a woman). You'll even pick up a batch of simple, effective compound exercises you can do at home.

- **Bones (Chapter 4)** strips you down to your skeletal framework. You'll look at the miniature factories that work inside your bones and learn how to preserve your spine with beneficial standing, sitting, and sleeping postures.

- **The Doors of Perception (Chapter 5)** shows you how your sense organs—eyes, ears, nose, and tongue—gather information about the outside world. You'll learn why nearly everyone eventually becomes nearsighted, what to do about ear wax, and how to tell if you're a "supertaster."

- **Your Lungs (Chapter 6)** sucks you into the world of your respiratory system. You'll consider unsettling topics that range from snoring to nasal mucus, examine the fine particles that lodge in your lungs, and get practical techniques for deep breathing and voice projection.

- **Your Heart (Chapter 7)** introduces you to the pump inside. You'll experience a virtual heart attack, learn the essential facts about blood pressure, and see how heart-rate monitoring can help you shape the perfect cardio workout.

- **Your Digestive System (Chapter 8)** leads you on a tour through the winding passages of your digestive tract—a nearly 30-foot-long canal that travels from tongue to toilet. Along the way, you'll sort out the probiotics from the prebiotics, make friends with fiber, and learn what your stool tells the world about you.

- **Your Immune System (Chapter 9)** takes you to the battleground between your body and its many enemies. You'll meet the bacteria that colonize your digestive system, greet the viruses that hijack your body, and discover what turns ordinary cells into cancer.

- **Sex and Reproduction (Chapter 10)** pulls back the curtain on the male and female sexual organs, explaining how they work and what trips them up. You'll also learn what health class didn't tell you about fertility, female pleasure, and the male obsession with size.

- **Your Final Exit: Aging and Death (Chapter 11)** takes you to the end of your life, and gives you a long, hard look at what it holds. You'll learn why you age, why you die, and what's most likely to do you in. Finally, you'll get a sobering picture of your final moments, complete with a side trip to the bizarre world of a near-death experience.

Separating Truth from Speculation

The science of human biology evolves rapidly, and the insights in this book are based on many of science's most recent discoveries. However, as with all scientific knowledge, it's always possible that better, more comprehensive studies will overturn the concepts we use today or change the way we think about them. In fact, it's a given.

When dealing with scientific research, this book won't bury you in footnotes. (The basic idea is that footnotes are only as good as the research on which they're based, and it's easy to cite a great deal of nonsense written by a great many people.) Instead, this book uses careful wording to distinguish rock-solid truths from tantalizing speculations. When this book says, "some scientists believe," you're about to meet a promising new idea that has some heavyweight science behind it, but hasn't convinced everyone. When this book says, "one study found," you're looking at some provocative new evidence that's on the cutting edge of scientific research.

About the Missing CD

This book is designed to give you a better understanding of how your body works. As you read through the chapters, you'll find references to books and websites that offer additional info, insights into current research, and more. To find a concise, chapter-by-chapter list of all these books and sites, head to the Missing Manuals home page (*www.missingmanuals.com*) and click the Missing CD link; then scroll down to *Your Body: The Missing Manual* and click Missing CD.

You can also find updates to this book on the Missing CD page by clicking the "View Errata for this book" link at the top of the page.

You're invited and encouraged to submit corrections and updates for this book, too, by clicking the "Submit your own Errata" link on the Missing CD page. To keep the book as up-to-date and accurate as possible, each time we print more copies, we'll include any confirmed corrections you've suggested. We'll also note all the changes to the book on the Missing CD page, so you can mark important corrections in your own copy of the book, if you like.

About Missing Manuals.com

To see the latest Missing Manuals videos, the most recent blog posts by Missing Manuals authors, the most recent community tweets about Missing Manuals, and special offers on Missing Manuals books, go to the Missing Manuals home page (*www.missingmanuals.com*).

We'd love to hear your suggestions for new books in the Missing Manual line. There's a place for that on the website, too.

And while you're online, you can register this book at *www.oreilly.com* (you can go directly to the registration page at *http://tinyurl.com/yo82k3*). Registering means we can send you updates about this book, and you'll be eligible for special offers, like discounts on future editions of *Your Body: The Missing Manual*.

You might also want to visit O'Reilly's feedback page (*http://missingmanuals. com/feedback.html*), where you can get expert answers to questions that come to you while reading this book. You can also write a book review on this page, as well as find groups for folks who share your interest in *Your Body: The Missing Manual*.

Safari® Books Online

When you see a Safari® Books Online icon on the cover of your favorite technology book, it means the book is available online through the O'Reilly Network Safari Bookshelf.

Safari offers a solution that's better than e-Books. It's a virtual library that lets you easily search thousands of top tech books, cut and paste code samples, download chapters, and find quick answers when you need the most accurate, current information. Try it free at *http://my.safaribooksonline.com*.

1 Skin: Your Outer Layer

Every practical person knows the value of a good outer layer. That's why you wear a raincoat when you visit London, use plastic wrap to save today's dinner for tomorrow's lunch, and don a ski mask when you climb snow-capped mountains (or make an unscheduled withdrawal from someone else's bank account). But by far, the most impressive covering you'll ever encounter is your *skin*—the 8 to 11 pounds of watertight wrapping that covers virtually every square inch of your body.

Skin does far more than the obvious task of keeping your insides on the inside. It's a washable, stretchable, self-repairing fabric that lasts a lifetime with minimal care. It's also home to a few other important bits of human machinery, including your hair, nails, and sweat glands. Removed and laid flat, your skin occupies about 20 square feet of space—enough to cover the top of a twin-size bed and make it the surprise winner of the "largest organ in your body" award.

Learning about your skin is a great way to spend a Sunday afternoon (and a surefire way to impress your dermatologist). That's because the study of skin holds secrets that can help you smell nicer (page 27), stave off wrinkles (page 19), and commit the perfect crime (see the box about fingerprints on page 27). In this chapter, you'll learn everything you need to know to care for your very own body wrapper.

The Protective Wrapper

When people think about the purpose of skin, most settle on the obvious—the way a few millimeters of tissue keeps their blood from oozing messily out of their body.

While a bit of skin certainly helps hold you together, it also plays several additional roles. First and foremost, it's a *protective barrier* that separates you from the harsh world outside. It helps keep water and nutrients inside your body, where they belong, and it keeps undesirable elements—like toxins and marauding bacteria—outside.

The Practical Side of Body Science

Skin Care

It might make you a little queasy, but everything you do to care for your skin—slathering on moisturizer, scrubbing with a sponge, and so on—you do to a layer of lifeless cells. Your morning shower involves scrubbing away the oldest and loosest skin cells—not to reveal the living cells underneath (which aren't tough enough to face the outside world)—but to reveal more dead skin. In this respect, people are rather like trees, covered in a dead-as-a-doornail layer of protective bark.

But don't give up on your skin just yet. Dead as it may be, your skin cells still need proper upkeep. Here are some points to consider:

- **Basic cleaning.** Neatniks take heart—even dead skin needs a regular bath. If you leave your dead skin undisturbed, it will mix with sweat and dirt to form a very tasty snack for the bacteria that live on your skin. As the bacteria digest this mixture, they produce a foul smell that will earn you some extra personal space on the subway.

- **Moisturizing.** Ordinary soaps are harsh and drying. They strip away the natural oils in your skin. Unfortunately, this dry skin loses its natural protection against bacteria, which can then slip in through cracks and fissures in your skin. To keep your defenses up, rub lotion on your hands when they become dry (for many people, that means after every washing), and use gentle cleansers on other parts of your body (like your face).

- **Exfoliation.** Some people swear by special scrubs and brushes for removing dead skin cells. While exfoliation may improve the feel of your skin and temporarily enhance its appearance, exfoliation overachievers are likely to end up with dry, inflamed skin. So if you're an exfoliating junkie, limit your sessions to twice a week, and moisturize your skin to replace the natural oils you've just scrubbed away.

Building a Barrier

To understand how your skin works its defensive mojo, you first need to understand that it's actually made up of two distinct layers: the *epidermis* (which is on the very outside) and the *dermis* (which is just underneath the epidermis).

The epidermis is your body's first line of defense. It transforms dead skin cells into a tough, protective layer.

> **Note** *Cells* are the smallest building block of life. All living creatures—from slimy amoebas to still slimier car salesmen—are made up of cells. Your body contains trillions of cells, many of which don't belong to you at all. (In fact, the teeny bacteria that digest food in your intestines account for more than half of the cells in your body, as you'll learn on page 196.)

Healthy skin cells start at the bottom of your epidermis, about ⅓ of an inch down, living an easy life and cheerily reproducing. As these cells mature, they get ready to face the outside world by producing a fibrous, water-proof compound called *keratin*. Keratin is a biological wonder substance. Your body uses it to build your nails and hair, and it's the basis of some of the sexier trimmings of other animals, including claws, horns, hooves, scales, shells, and beaks.

When your body produces fresh skin cells, these newcomers push the older cells out of the crowded neighborhood at the base of the epidermis and toward the surface of the skin. The trip takes anywhere from a couple of weeks to a month. By the time a skin cell reaches the surface, it's little more than a dead, scale-like structure that's filled with keratin but none of the ordinary cellular machinery. Each surface skin cell lasts about 30 days on the outside, which means you get an entirely new skin every month.

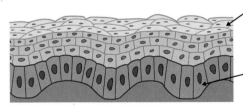

The top layer of your skin. It's not such a good neighborhood, as everyone's dead.

The bottom layer of the epidermis, where healthy skin cells are born.

On most of your body, the epidermis is barely thicker than this page. However, the skin on the palms of your hands and the soles of your feet is much thicker, so it can spend all day slapping up against the outside world without wearing off.

Shedding Your Skin

Every day, you lose millions of dead skin cells. They don't fall off all at once—instead, you leave a trail of shed skin everywhere you go. We could tell you how many you lose each minute, but it's really not that important and likely to make you a little nauseous. (All right, if you insist—30,000 or so scales of skin flake off your body every minute. Right now, they're collecting on the pages of this book, on your clothes, on whatever piece of furniture you're sitting on, and so on. Over the course of a year, you lose about a pound of the stuff.)

You might wonder why you never see much of this skin lying around. That's because once your skin leaves your body, it's known by another name: *dust*. Good estimates suggest that the majority of the material you vacuum off your carpet every week (or every month, or every year) are errant skin flakes. That means that when you clean your house, you're vacuuming up bits and pieces of yourself and the people who live around you. Yes, there's some genuine sock lint in there, some cookie crumbs, and a bit of tracked-in-from-outside dirt, but it's mostly skin. Because skin flakes are thin and nearly transparent, your household dust almost always has a light, silvery-grey color.

Note If you want to take a look at your dead skin before it ends up somewhere else, you can try this somewhat unsettling experiment: Stick a piece of clear tape on the back of your hand, strip it off, and then hold it up to a light. You'll find hundreds of freshly shed skin cells preserved for your inspection.

The Creature That Eats Your Skin

It turns out that your skin flakes have yet another name: *lunch*. That's what they are to an unusual family of creatures that exists on a diet made up entirely of dead skin. (And no, they're not zombies.)

The culprits are *dust mites*—very tiny, distant relatives of the common household spider. Dust mites live in our houses by the millions, with most of them taking up residence in upholstered furniture, drapery, carpets, and—above all—mattresses. Dust mites need just three things for a life of contentment: warmth, moisture, and a steady diet of skin flakes. In your bed, they get all three.

You won't actually see the dust mites that share your home, because they're vanishingly small (a family of mites could pack themselves into the period at the end of this sentence). But if you looked at one under a microscope, you'd see an otherworldly, eight-legged creature.

> **Note** Before you let the idea of dust mites ruin your day, remind yourself that, unlike some other mites and other nasties, dust mites don't actually live *on* your skin—they live in the fabric of the objects around you. In fact, dust mites have absolutely no interest in crawling on your body.

If you're like most people, dust mites are no big deal and you can safely forget about them. But for some people (estimates suggest one to three people out of 10), dust mites can trigger allergies and even asthma attacks. Common symptoms of dust-mite allergies include sore eyes, an itchy throat, and sneezing fits. If you think you might be allergic to dust mites, it's worth going to an allergy specialist, who can give you a quick and painless skin-prick test. If you *are* allergic, you may want to use some of the tips in the box on the next page to help reduce your symptoms.

The problem isn't the mites themselves—it's their excrement and (ironically enough) the skin they shed. And here's more information you probably don't want to know: Dust mites actually eat and excrete the same skin flake several times, until they've finally digested all the goodness out of it.

> **Note** As far as critters you don't want to think about go, there's good news, too: Two stubborn skin dwellers that have plagued humankind for generations—the human flea and the body louse—are no longer much to worry about. In Elizabethan times, these creatures crawled into bed with virtually everyone, rich and poor. Today, thanks to relatively simple conveniences like scalding-hot water and laundry machines, these pests (and the unrelenting itchiness they cause) are virtually unknown in the Western world.

The Practical Side of Body Science

Controlling the Mite Population

If you have dust mite allergies or asthma, you may be able to improve your life with a bit of extra work. These tips can help cut down on the number of dust mite colonies that live with you:

- **Control dust.** Vacuum often, dust flat surfaces, switch from carpet to hardwood floors, and remove knickknacks that collect dust. None of these steps will kill dust mites, but you can keep their numbers down by reducing their food supply.

- **Control humidity.** Dust mites thrive in moist environments. Sadly, no matter how dry your house is, your breathing and perspiration provides more than enough dampness to keep them happy in your bedding.

- **Use cold and heat.** If you can wash your bedding at scaldingly high temperatures—at least 130 degrees Farenheit—you can kill the mites that are there (although this obviously has no effect on the many more mites in your mattress). If you have a plushy object you can't launder, like a child's stuffed toy, a day in the freezer will also kill the mites, although it may leave lint on your frozen peas.

- **Use allergenic covers.** Many companies sell zippered covers for mattresses and pillows that can reduce the number of dust mites that get into your bedding and the amount of allergenic excrement that comes floating out once they're in it. Of course, some mattress covers are about as comfortable as sleeping on a vinyl tablecloth. And frequent laundering may stretch the microscopic pores of the cover so that they're big enough to let everything through anyway. If you decide to try this approach, it's worth doing some research before you buy.

- **When travelling, don't think about it.** Sure, there are probably plenty of dust mites in hotels, bed-and-breakfasts, and so on, but you'll be home soon enough. If you really must feed your paranoia, obsess about something more serious, like bed bugs (see *http://en.wikipedia.org/wiki/Bedbug* for travel tips that can help you spot these very unwelcome bedmates).

Skin and the Sun

The thought of voracious mites crawling through your bed linens and devouring your dead skin is an unpleasant one. However, the real danger to skin lies in something that seems a lot more innocent—leisurely summer days and the warming rays of the sun.

Your skin has a bit of a love-hate relationship with the sun. On one hand, the sun fuels miniature chemical factories in your skin that create *vitamin D*, a key nutrient for your body. On the other hand, the powerful rays of the sun damage skin cells, occasionally scrambling their genetic material enough to trigger deadly cancers. In the following sections, you'll find out what you can do about it.

Vitamin D

Not long ago, vitamin D was considered dull and definitely unsexy. Sure, it was known to help your body absorb calcium and prevent *rickets* (a childhood disease that softens the bones and causes debilitating deformities). But adding a dash of vitamin D to milk and a few other vitamin-fortified foods solved the problem, and no one thought much about vitamin D—until recently.

Today, vitamin D has leapt to the forefront of the supplement world, thanks to several new studies that suggest it plays a role in the prevention of cancer and other diseases. It's no longer treated as a simple calcium-booster—vitamin D now has its own starring role as a *hormone* that triggers a range of cellular processes. Time will tell if science validates this promising new research, or if it becomes another dead end in the vast maze of nutrition science. In the meantime, there's good reason to make sure your body has a solid dose of the stuff.

Vitamin D is naturally present in very few foods, but your skin has the ability to create this wonder drug when you expose it to the ultraviolet rays of the sun. The cells that carry out this operation lie at the bottom of your epidermis. You need surprisingly little exposure to the sun to maintain a healthy supply of vitamin D. The rule of thumb is 10 or 15 minutes of direct sun exposure, two or three times a week, on just part of your body (say, your face, hands, and arms). After that, it's time to reach for the sunscreen.

Unfortunately, the vitamin D manufacturing process doesn't work well in diffuse sunlight—say, in the winter months of a Northern state. Cloud cover and pollution also dramatically reduce the amount of ultraviolet light that reaches your skin. For example, in Boston, sunlight is too weak to trigger vitamin D synthesis from November through February. To make up the difference, you can take a vitamin D supplement—typically, 1,000 IU each day (look for this measure on the bottle), until summer rolls around again. This is roughly the amount of vitamin D that you'd get from 10 glasses of milk.

Supplementing your diet with vitamin D is particularly important if you have brown or black skin, because this natural sunscreen makes it more difficult to synthesize vitamin D.

> **Note** The key point to remember is that the amount of sun exposure you need to synthesize vitamin D is very little in the summer months (or in a tropical climate). But in late fall and winter, you can run around in boxer shorts without producing a microgram of vitamin D.

Sun Damage

So far, you've heard about the good side of the sun—its ability to fuel your skin's vitamin D factory (at least in warmer seasons). But here's the scary part: The dangers of the sun far outweigh its benefits.

The problem, of course, is *skin cancer*. Skin cancer is by far the most common form of cancer, trouncing lung, breast, prostate, and colorectal cancers. However, many skin cancers disfigure the skin without threatening your life. Only the type of skin cancer called *melanoma* is likely to spread from your skin to the rest of your body, which it can do quite quickly.

The culprit is the *ultraviolet radiation* in sunlight. Ultraviolet rays can be divided into three types: UVA, UVB, and UVC, from least worrisome to most dangerous. *UVA* can age the skin and may play a role in skin cancer, but science still considers it to be the least harmful. *UVB* is the type of radiation that fires up vitamin D production and triggers sunburns (and eventually skin cancer). *UVC* is more dangerous still, but in most parts of the world it's blocked by the ozone layer, which means it never reaches your skin.

UV exposure increases your risk for all types of skin cancer. But like many things that have adverse health consequences (smoking, obesity, and so on), there's a lag between the behavior and its effect. In the case of skin cancer, the blistering sunburn you get in your early twenties might lead to skin cancer 30 or 40 years later. Furthermore, the effects of sun exposure are cumulative, so it may take many years of sun-inflicted damage before you harm some of your skin's genetic material beyond repair.

Most dermatologists believe that skin cancer is highly preventable if you practice good sun habits:

- **Reduce sun exposure.** Don't linger in the sun between 10 a.m. and 3 p.m. (11 a.m. and 4 p.m. during daylight-saving time). If you find yourself in strong sun, seek shade.

- **Cover up.** Always wear a wide-brimmed hat, long-sleeved shirt, and long pants on sunny days.

- **Use sunscreen.** Look for a product that protects against both UVA and UVB rays and offers a sun protection factor (SPF) of at least 15. Applied properly, a sunscreen of SPF 15 protects the skin from 93% of UVB radiation. Higher SPF numbers are better and may block more UVA, but the difference is not nearly as significant as the numbers imply.

- **Use sunscreen *properly*.** Apply sunscreen 15 to 30 minutes before going out. Repeat every two or three hours, more often if you're swimming.

Note Despite decades of use and study, the science of sunscreens isn't settled. Although sunscreens clearly reduce the occurrence of less harmful types of skin cancer, several studies have found that they offer no protection from deadly melanoma. The reason for this discrepancy is unknown—some experts believe sunscreen gives people a false sense of security, allowing them to stay out longer in potentially harmful sun. Others believe the culprit is not using enough sunscreen or not applying it properly, while still others blame old sunscreen formulations that failed to block UVA rays and contained now-banned ingredients. The best advice is to use sunscreen in conjunction with *all* the good advice in this list.

- **Avoid tanning beds.** Although tanning beds use UVA rays rather than the more damaging UVBs, they're far from harmless. Even occasional tanning sessions accelerate skin aging and are likely to increase your risk of skin cancer.

- **Learn from your mistakes.** If you end up with a sunburn—even a mild one—figure out which rule you broke and resolve to avoid the risk next time. Remember: Sun damage accumulates over your lifetime.

These guidelines are particularly important if you have light skin, a large number of moles, or a family history of melanoma, all of which single you out for greater risk.

Note Scientists believe that skin exposure is particularly risky for children. Each severe sunburn before the age of 18 ratchets up the risk that skin cancer will develop later in life. So make an extra effort to follow these rules with children and teenagers, keep babies under one year old out of direct sun in the summer, and never leave infants playing or napping in the sun.

Tans, Moles, and Melanoma

Excessive sun exposure and blistering sunburns are clearly a bad idea. But is there anything to fear from the healthy glow of a modest tan?

According to current-day science, the answer is "probably." Biologically speaking, tanning is what happens when ultraviolet-light exposure causes your skin to start producing more *melanin*. Melanin is a pigment found deep in the skin, at the base of the epidermis. It's responsible for freckles, moles, and beauty marks—which are all more or less the same thing—and for the differences in human skin color.

But here's the bad news. Your tan is a defense mechanism that responds to damage caused by the sun. As ultraviolet light strikes your skin, it shatters the DNA in some unlucky cells. Your body has ways of catching and repairing this sort of damage, but when the rate of damage escalates, it presses the panic button and triggers a tan. In other words, the process that prompts melanin production and tanning is the same process that causes sunburns and may, eventually, cause skin cancer.

Late-Night Deep Thoughts

How the Sun Created Race (and Kinky Hair)

The paradox between the positive and negative effects of the sun is at the root of humanity's wide range of skin colors.

Races that have traditionally lived in the glare of the equatorial sun have darker skin that offers natural sun protection. But here's the curious bit: If you follow the tree of humanity back far enough, you'll find that we all have relatively dark-skinned ancestors. The light-skinned people of today descended from mostly European races that lost their built-in sunscreen over generations of life in colder, dimmer lands. The obvious question is *why*. After all, if the sun is so dangerous, surely everyone can use a bit of natural sunblock.

The going theory is that when dark-skinned people moved north and south, their natural sunscreen starved them of vitamin D. But the genetic misfits with light skin absorbed more sun, providing the vitamin D they needed, and they thrived.

The sun played the same role in shaping humanity's hair. Races that needed sun protection developed UV-scattering Afro hair to shield their scalps. Races that needed more vitamin D developed straight, kink-free hair that allows light to pass through to the skin below.

In other words, if the sun wasn't harmful, no one would sport dark skin. And if vitamin D wasn't important, there would be no white people.

Finally, it's worth being on the lookout for melanoma, particularly if you're at high risk. A melanoma begins life looking like a harmless mole. However, a few red flags suggest the possibility of a problem. If you notice any of the signs below (known as the "ABCDE rule"), see your doctor and get checked. If you catch a melanoma developing in your skin before it goes too deep, the chance of successful treatment is high.

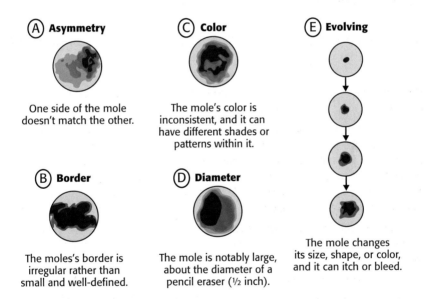

(A) **Asymmetry**

One side of the mole doesn't match the other.

(B) **Border**

The moles's border is irregular rather than small and well-defined.

(C) **Color**

The mole's color is inconsistent, and it can have different shades or patterns within it.

(D) **Diameter**

The mole is notably large, about the diameter of a pencil eraser (½ inch).

(E) **Evolving**

The mole changes its size, shape, or color, and it can itch or bleed.

If you have a large number of moles on your skin, it's important to keep track of your collection. That way you'll notice when a new mole appears or an existing one changes. Your doctor can help by taking yearly pictures.

Going Deeper

So far, you've focused on the epidermis, which is the mostly dead, top layer of your skin. Under the epidermis is a thicker second layer of skin with a whole lot more going on. This layer is called the *dermis*.

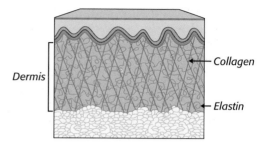

The dermis shapes and supports your skin with tough connective tissue that uses strong, flexible fibers made of *collagen* and *elastin*. These fibers make your skin stretchable and resilient so you don't tear a hole in your armpit when you reach for a box of macaroni on the top shelf. These fibers also keep your skin toned and wrinkle-free through the first half of your life, after which they start to loosen and unravel.

Underneath the dermis is a layer of tissue that isn't part of your skin. Known as the *hypodermis*, it binds your skin to your body and stores the sorry globs of fat that you'll consider in the next chapter.

Unlike the epidermis, which is constantly dying off and renewing itself, your dermis is yours for life. A dense network of blood vessels supplies your dermis with essential nutrients. If you have an open cut, immune cells in the dermis fight infectious evildoers that try to creep in (a process you'll learn about in Chapter 9).

> **Note** Tattoo machines work by puncturing the skin thousands of times and inserting small globules of ink. The reason that tattoo ink doesn't wash off or flake away is because it's not in the epidermis—instead, it sits near the top of your permanent dermis layer. However, the fine detail in a tattoo will fade with time, because the ink droplets drift slightly.

The dermis is also home to many more important bits of body hardware. For example, every square inch of skin has thousands of buried nerve endings that react to different types of sensations, including cold, heat, pain, and pressure. The dermis also holds oil glands, sweat glands, and hair follicles—three key players you'll learn about in the rest of this chapter. But first, it's time to consider what happens as your skin ages, loosens, and gives way to *wrinkles*.

Troubleshooting Your Skin

Now that you understand the two layers of skin that wrap your body, you can make sense of a whole host of skin insults and injuries.

- **Cuts.** Shallow cuts that don't dig past the epidermis won't produce any blood. Instead, your skin will just ooze clear liquid. If you see blood, you've hit the much thicker and much tougher dermis.

- **Stretch marks.** Extreme skin stretching (for example, stretching caused by extreme weight gain or pregnancy), can tear the dermis, leaving stretch marks behind. Although they fade with time and are completely harmless, stretch marks never disappear. (Incidentally, 75 to 90 percent of women get some sort of stretch marks during pregnancy. Special skin creams seem to offer little help, and the natural elasticity of your skin appears to determine how severely you're affected.)

- **Blisters.** A blister is a fluid-filled pocket that forms between the epidermis and the dermis. The culprit is usually friction from a repetitive motion. Blisters heal fastest (and with no chance of infection) if you leave them undisturbed and unbroken.

- **Bruises.** A bruise is an injury that causes blood to seep from damaged tissue into the dermis. The bruise remains until your body reabsorbs the blood.

- **Warts.** A wart is a small, rough skin tumor that's usually triggered by a virus (and you usually pick up that virus from a damp surface, like a community swimming pool). Warts are more likely to cause pain and embarrassment than any serious complications. However, they can be tenacious, especially those that appear on the undersides of your feet. If you can't kill off a wart with the standard, over-the-counter products, visit your doctor for a more powerful approach, like a dab of super-cold liquid nitrogen.

Wrinkles

As you age, the style of your skin changes. You start with a tight-fitting sports jacket, and you wind up with something closer to a pair of baggy pajamas. This transition is quite traumatic for many people, as our culture considers it deeply embarrassing for one's body to betray any sign that it's a day over 18. If given the choice to look wise and experienced or young and nubile, most of us would choose the baby face every time.

Many factors work together to cause midlife wrinkles and the pruniness of old age. As the years tick by, the collagen and elastin fibers in your dermis—those components that make your skin flexible and resilient—begin to break down, loosening their hold on your skin. Unfortunately, there's not much you can do to intervene. But if you must try, here are a few wrinkle-avoiding strategies:

- **Choose the right parents.** Your genes have the greatest say in deciding how elastic your skin is, and how long it stays relatively smooth and unwrinkled. That's why some people in their sixties look like they're in their thirties, much to the chagrin of everyone around them. If you want a quick prediction of how your skin will fare over the next few decades, look at your parents. And if this leaves you too depressed to continue through the rest of this chapter, consider the possibility that you were adopted from a passing circus.

- **Don't use your face.** Many of the deeper grooves in your face are usage lines that mark where your skin folds when you scowl, smile, frown, or look utterly confused. To reduce the rate at which these wrinkles form, stop expressing any of these emotions. Or just accept the fact that wrinkles add character to your face.

> **Note** A variation of this wrinkle-avoiding technique is Botox injections, which paralyze the face muscles using a highly toxic nerve agent. (It's the same substance that causes death by paralysis in improperly canned foods.) A Botoxed face temporarily loses some of its ability to move, and a face that can't move has a hard time furrowing up a decent wrinkle. What you get is a sort of blander, wax museum version of your face. If you prefer being wrinkle-free to being able to move your forehead, Botox just might be your ticket.

- **Don't smoke.** Cigarette smoke damages skin, causing it to wrinkle prematurely. This probably happens because cigarette smoke reduces blood flow to your skin, starving it of important nutrients. And while quitting the habit may improve your lungs, it won't repair skin that's already sagging.

- **Limit sun exposure.** Ultraviolet light (both UVA and UVB) breaks down the collagen in your skin. This weathering process speeds up aging and increases wrinkles. To prevent sun damage, slap on some sunscreen and follow the good sun habits described on page 15.

These techniques may slow the rate at which your skin becomes progressively more wrinkled, but what can you do to remove the wrinkles you already have?

There's certainly no shortage of cosmetic products that promise age-conquering miracles. However, most skin creams do relatively little. On the practical side, they may moisturize your skin (as dry skin looks older) and shield it from sun damage (with sunscreen). The effect of other ingredients is less clear-cut. Although many anti-aging skin creams are packed with anti-inflammatory ingredients, their concentrations are low and there's little independent research to suggest that they actually do anything. Similarly, vitamins, collagen, antioxidants, and other useful-sounding substances are unlikely ever to reach the lower-level dermis, which is where wrinkling takes place. Some creams contain ingredients that obscure fine wrinkles or scatter light, giving skin a "soft-focus" effect. Whatever the case, these creams can only hide aging rather than make lasting improvements. And lotion lovers beware: Some ingredients can actually aggravate sensitive skin or clog pores, exacerbating acne (page 22).

More drastically, cosmetic procedures like chemical peels, laser resurfacing, and microdermabrasion can improve wrinkles by removing excess dead skin in a strategic way. The effect is temporary, usually limited to fine lines rather than deep wrinkles, and may cause redness and peeling. For all but the most wrinkle-averse, it hardly seems worth the trouble.

The truth is that if you live in your body for half a century, it will gradually develop the creases of use and abuse. The over-80 crowd will tell you that the relentless march of time leaves the human face with more grooves than a 45-rpm record (but first they'll have to explain what a 45-rpm record is). The real decision you have to make is not how to fight wrinkles, but whether you want to accept them with dignity or become an increasingly desperate chaser of youth.

The Oil Factory

Human skin has two types of glands, and both are happily squirting their stuff onto your skin right now.

Most obvious are the sweat glands, which dampen your skin to help you cool down. (You'll learn about these on page 25.) However, your skin also has a set of oil glands, which are properly called *sebaceous glands*. Most sebaceous glands wrap around one of the millions of fine hairs that line your body. When your body kicks them into action, they release an oily substance called *sebum*, which oozes down the hair follicle and out onto your skin.

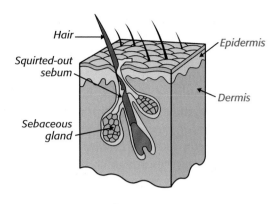

Although it may not be the most suitable dinnertime topic, sebum does a lot for your body. It contains a mixture of substances that softens your hair, moisturizes your skin, and discourages bacteria growth. (In some water-loving animals, sebaceous glands help waterproof fur.)

Note Many skin lotions soothe and moisturize the skin using *lanolin*, which is the greasy yellow sebum produced by domestic sheep. In fact, the product name *Oil of Olay* is derived from rearranging some of the letters in the word "lanolin" (and adding a few more). So the next time you visit a farm, don't hesitate to rub your face against a sweaty sheep—it could do wonders for your skin.

Acne

Less helpfully, sebaceous glands cause *acne*, the scourge of teenagers everywhere. The problem starts when puberty ramps up the production of certain hormones, most significantly, **testosterone** (in both boys and girls). At the first sign of these hormones, the highly excitable sebaceous glands begin pumping out huge quantities of sticky sebum. Inevitably, they clog themselves up. But the real nightmare is that they keep producing sebum even when the glands are blocked, causing a swelling that eventually appears on the surface of the skin as a *whitehead*. With chronic acne, swollen sebaceous glands become inflamed, and trapped sebum can form a cyst. Cysts, in turn, can lead to permanent scarring.

So a blockage deep in a sebaceous gland causes acne, which itself is usually caused by the sudden onrush of hormones at puberty. It's just as important to note what *doesn't* cause acne, including chocolate, fried foods, and poor hygiene. (In fact, aggressive washing can exacerbate the inflammation.) Stress may make acne worse, which is rather unfair, considering that the stress was probably caused by the monster zit that appeared Saturday night before your big date.

If you're suffering from acne, here's some practical advice:

- **Think before you squeeze.** Virtually all dermatologists will tell you to resist the urge to pop a zit. After all, the risks are legion—you might force the sebum deeper into your skin, worsen the inflammation, and cause scarring. However (and this is not the best topic among polite society), if you have a pimple that isn't inflamed and is white, ripe, and raised above the surface of your skin, it's safe to give it a tentative nudge. But if blood or clear liquid emerges, just walk away from the mirror before you do any damage.

- **Try an over-the-counter lotion.** Treat mild cases of acne with an over-the-counter cream. The key ingredient to look for is **benzoyl peroxide**. There's no magic formula, so don't plunk down serious cash for the miracle cures shilled on late-night infomercials.

- **Get help for acne that doesn't improve.** Don't be an acne hero. Living with a bad case of acne can lead to scarring (and not just the psychological kind). Your friendly neighborhood dermatologist can prescribe an antibiotic lotion or an oral antibiotic that will change the balance of bacteria on your skin, ultimately reducing the inflammation.

Your Temperature Control System

In some respects, the life of a reptile has a lot of appeal. When the sun rises on a Monday morning, springing out of bed is the last thing on any lizard's mind. Much as you may need a second or third cup of coffee before you can string a coherent sentence together, a lizard can't do much of anything until it's spent a long, lazy morning basking in the sun, heating its body to operating temperature.

Warm-blooded humans like you don't work that way. Your internal temperature stays at a balmy 98.6 degrees Fahrenheit (or thereabouts). This is quite a feat, because your body continuously generates heat—primarily by your muscles contracting during routine activity and by major organs like the liver. To cool down, your body needs to release some of that heat into the air around you.

Using a design that's the human equivalent of a hot-water radiator, your body sends warm blood to the surface of your skin so it can radiate heat away to the cooler world outside. When you need to conserve heat, your body clamps down on this process, tightening the blood vessels in your skin. That reduces the flow of blood near the skin and slows your rate of heat loss.

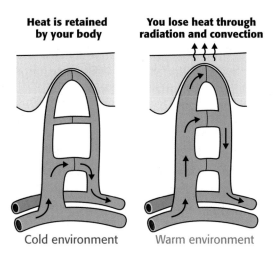

Heat is retained by your body

You lose heat through radiation and convection

Cold environment   Warm environment

This system explains why people become flushed when they're hot (it's from the increased blood flow). It also explains how *frostbite* inflicts its damage. The cold itself doesn't harm your body—instead, the extremely reduced blood flow starves your cells of the oxygen they need to survive.

Frequently Asked Question

How Much Heat Do You Lose Through Your Head?

It's an often-repeated, slightly wonky story. Cover your head on a cold day, because 40 percent of your body heat exits through your cranium. Or 60 percent. Or 80 percent. Sure, the explanations are a bit dubious—the skin on your scalp is extremely thin, heat rises, and if you're wearing clothes the heat has nowhere else to go—but who can question such an enduring yarn?

Serious-minded scientists have pointed out that if the 60 percent figure were true, you'd be more comfortable on an Alaskan cruise with nothing on but a ski hat than if you were fully dressed but bare-headed. So perhaps it's no surprise to find that the true figure is somewhat less than 10 percent. In fact, you lose little more heat out of your head than you lose from any similarly sized part of your body, although your face, head, and chest are more sensitive to temperature changes, which may give you the impression that you're colder.

This confusion might have resulted from a flawed interpretation of a military study that examined heat loss in fully dressed soldiers. The soldiers were bundled up in survival suits, but hatless, so their heads did account for about half of the heat they lost. (If they war-gamed naked, the equation would change.) In any case, a hat remains a sensible addition to any cold-weather ensemble.

Blushing

One thing this system doesn't explain is the uniquely human habit of *blushing*, in which sudden embarrassment causes increased blood flow and pronounced reddening, particularly in the face. Scientists guess that blushing may be an involuntary skin signal designed to solve social problems. It works like this: If you get into a sticky situation with a more dominant member of your social tribe, blushing expresses your remorse and gets you off the hook without the need for physical violence. Experts agree that the best way to deal with blushing is to announce it and accept it (for example, by saying something along the lines of, "Oh drat, I'm about to blush again!"). Trying to hide it usually triggers a cycle of increased embarrassment and increased blushing, turning the skin of a sensitive person to a distinct shade of cranberry jelly.

Sweat

The body's heat-exchange system makes perfect sense, but on its own it's just not enough. Sure, your body can radiate heat through your skin, but on a hot day it won't lose a sufficient amount to keep you cool. To lose heat more efficiently, you need the help of *sweat*.

Sweat is part of your body's messy air-conditioning unit. Your body sweats continuously, but you don't notice the small amounts of moisture that trickle out because it's truly miniscule, and your body reabsorbs some of it. But when the outside temperature rises or the activity in your body soars (say, when you run to catch the last bus home), your body ramps up its sweat production.

> **Note** Stress also causes sweating. Other than the obvious purpose (to embarrass you in your third-grade public-speaking competition), sweating in response to stress works as part of your body's *fight-or-flight response*. Essentially, your body assumes that you're either going to run away from or attack the threat in front of you, so it prepares for the imminent increase in body heat by switching on your natural air conditioning.

Sweat is mostly water, with a pinch of salt and tiny amounts of other waste products thrown in. As sweat evaporates, it takes some of the heat from your skin, noticeably cooling it. (And if you don't think it's noticeable, try taking a hot shower and then walk around the house without drying yourself.)

But the real point of sweat isn't to cool your skin, but to cool your *blood,* thereby maintaining your internal body temperature. To accomplish this, your body uses the blood redirection trick you saw on page 24. When you sweat, your body sends more blood to the newly cooled surface of your skin. The blood gets a chance to cool down, and then it gets pumped back deeper into your body. This isn't all that different from the way a refrigerator works—it circulates a special substance (ammonia gas) through coils at the back. Once this substance cools, it's returned to the inside of the fridge so it can keep your rutabagas fresh.

> **Note** Say what you like about farm animals and zoo dwellers, but humans are the undisputed sweating champions of the natural world. In fact, many mammals barely sweat at all. Cats and dogs, for example, sweat only on their paws. (This is why dogs pant—they can't cool themselves sufficiently by sweating alone. The air they inhale cools the surface of their lungs and the blood that runs nearby.) Our habit of sweating probably explains why we don't have thick fur covering our bodies like some other animals—if we did, it would interfere with our ability to evaporate sweat from our skin.

Your skin is studded with several million sweat glands. They cover every square inch of your skin, with just a few exceptions (namely, your lips, nipples, and sexual equipment). The structure of a sweat gland is simple: It looks like a coiled tube that sits in the dermis (where your body produces sweat) and opens out through a pore. Some, but not all, sweat glands squirt their liquid out onto a hair, like your sebaceous glands do.

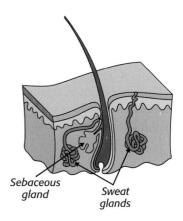

Sebaceous gland

Sweat glands

If you live in a cold or moderate climate, you can produce about one quart of sweat every hour. Move to the tropics and a few weeks later your body doubles or triples its maximum sweat-producing capacity. At the same time, your sweat becomes less salty.

How Fingerprints Work

Living on the wrong side of the law? If so, you'll want to spend some time thinking about *fingerprints*, the unique pattern of whorled ridges that adorns every human's fingertips.

Your fingerprints were formed, more or less at random, while you were still in your mother's womb. To biologists, fingerprints are known as *friction ridges*, and they're thought to improve our sense of touch. They might also give you a better grip on small, wet objects. And thanks to your sebaceous glands and your sweat glands, your fingerprints leave wet, oily tracks wherever they've been, which is of great interest to law-enforcement officers.

Incidentally, there's no shortage of exquisitely painful home-cooked approaches to alter or remove your friction ridges, including sandpaper, Super Glue, needles, and liquid nitrogen. Unsurprisingly, you can find all of these distinctly dimwitted ideas on the Internet. But before you try them out, consider investing in a pair of latex gloves instead. After all, when police interview suspects, they pay particular attention to the fellow with no fingerprints.

Body Odor

So far, we've skirted over one nagging question—namely, why does your personal air conditioner smell like dirty socks?

Surprisingly, sweat itself has no odor. You can douse yourself in the stuff without picking up the faintest scent. However, the bacteria that live on your skin aren't so innocent—they feed on your sweat and produce a rich collection of stinky substances. (This is the reason a discarded workout shirt smells worse the next day—the bacteria living on it have had some extra time to digest its tasty payload.)

As you've probably noticed, body odor seems to emanate from specific places in your body. To understand why, you need to recognize that your body actually has two types of sweat gland:

- **Eccrine.** These are the most numerous sweat glands. They're found all over your body and are particularly dense on the palms of your hands, the soles of your feet, and on your forehead. The eccrine glands do most of your body's temperature control.

- **Apocrine.** These sweat glands are concentrated in the forested areas—the armpits and genitals. Instead of secreting ordinary salt water, they squirt out a thick, milky fluid that has plenty of fats and proteins.

Apocrine glands almost always dump their contents onto a hair follicle. Unlike eccrine sweat glands, apocrine glands don't do much for temperature control, and they react more readily to emotions and sexual stimulation. Bacteria devour rich, apocrine sweat, leaving their signature gamey odors behind. Bacteria aren't nearly as interested in the watery sweat that leaks out of the eccrine glands, but under the right conditions they can still make a meal of it, along with skin oils and dead skin cells. (That's why a warm, moist, poorly ventilated foot can develop a room-clearing odor that rivals the sweatiest armpit.)

> **Note** Apocrine glands develop during puberty, which is why babies and toddlers don't have body odor problems.

This raises an excellent question—if apocrine glands don't help you cool your body, why are they there stinking up the place? It seems that the chief purpose of apocrine sweat is to create your distinctive body odor. Several studies have shown that women, when asked to smell a lineup of used undershirts, can pick out their man's shirt by smell. Thus, apocrine glands are the human equivalent of the sexual-scent glands of other animals— and whether they have a real effect or are just an evolutionary leftover depends on whom you ask.

Frequently Asked Question

Do Humans Have Pheromones?

In many other animals, smells are an important signaling mechanism that can indicate ownership and trigger mating. When chemicals have this effect, they're called *pheromones*. Essentially, pheromones act like hormones that travel out of the body. An individual secretes pheromones, they waft through the air, and they trigger a behavior in someone else.

Despite some top-flight scientific studies and many teenage fantasies, no one has ever discovered a chemical in humans that acts like a pheromone. However, there are plenty of tantalizing possibilities. Some provocative scientists suggest that pheronomes may underlie the mysterious chemistry of mate selection and explain why we tend to choose lovers who have distinctly different immune systems from our own. Others wonder if pheromones can explain why women living together may begin to menstruate on the same schedule. Both of these phenomena are highly disputed and might add up to nothing more than hot air. However, the possibility of mysterious chemicals controlling our destinies is thought-provoking. It's also enough to make you think twice before reaching for the deodorant stick.

Deodorants and Antiperspirants

When people learn how body odor works, the first question they usually ask is how they can stop it.

The first line of defense is bathing, which reduces your sweaty residue, leaving bacteria with a whole lot less to lunch on. However, soap and water won't kill the bacteria itself, which is a more-or-less permanent resident on your body. (You'll learn more about your skin-dwelling colonies of bacteria on page 214.)

Another popular tactic is to use deodorant or antiperspirant, which you usually apply to bacteria's favorite dining spot—the underarms. These two products work differently. *Deodorants* mask body odor (which should rightly be called *bacteria* odor) with a different smell. They may also contain powders that absorb moisture and chemicals that can kill some of the bacteria. Because you can never completely eradicate the bacteria, deodorant is really a population-control tactic.

Antiperspirants may include musky perfumes and germicides like deodorants, but they also have an aluminum-based chemical that temporarily blocks sweat glands. To be labeled an antiperspirant, clinical tests must show that the product actually works. This involves rather amusing studies that put a number of people in very hot rooms and get lab technicians to collect the resulting sweat. The rule of thumb is that a basic antiperspirant must reduce underarm sweating by 20 percent in most people. A high-powered antiperspirant (one with "maximum" or similar language on the label) must hit the 30-percent mark. Prescription antiperspirants can reduce sweating even more.

Now that you understand the science of your armpit, you're ready to learn about a significant drawback to antiperspirants: They only work on the comparatively harmless eccrine glands. So while antiperspirants do decrease the amount of wetness (which does slow down your armpit bacteria), they can't suppress the strong-smelling apocrine glands. So, after an hour at the gym, you'll still smell like, well, yourself.

Finally, it's important to address one of the very real risks of antiperspirants. No, it's not breast cancer or Alzheimer's disease, despite what you might have read in imaginative email chain letters. For the record, aluminum, the key ingredient in antiperspirants, is the third most common element on our planet, and it's found in food, air, and over-the-counter medications like antacids, all of which provide more aluminum than you can absorb from an antiperspirant through your skin. Furthermore, the amount of waste your sweat glands excrete is small, so there's no reason to think that slowing down a few sweat glands can increase the level of toxins in your blood.

The real danger of antiperspirants is *staining*. That's because the aluminum can react with your sweat to create an embarrassing yellowish stain on your favorite clothes. If this is a problem, apply your antiperspirant and walk around shirtless until it dries. Or consider switching to deodorant.

Note Deodorants and antiperspirants are simple ways to deal with ordinary sweat, but if you suffer from excessive sweating, you may need the help of the medical community. Your doctor can determine if your sweating is linked to another problem, such as thyroid disease, or if it's just genetic bad luck (in which case you have a range of treatment options, from stronger antiperspirants to underarm Botox injections and surgery). Lastly, look out for body-odor changes. For example, suddenly sweet body odor may hint at diabetes, or it could just be the result of a change in diet. If in doubt, have it checked.

Hair

Along with the different sweat glands and oil glands, there's one more type of body equipment rooted in the dermis—your hair.

Hair consists of long, flexible strands of dead cells. These cells are filled with *keratin*, the same wonder substance that strengthens the outer layers of your skin (page 9). The figure on the next page shows a hair erupting from the surface of the skin. As you can see, the surface of the *hair shaft* consists of overlapping scales, like shingles on a roof.

This close-up holds the secret to hair frizz. On a humid day, tiny water droplets work their way in between the scales of the hair shaft, making the hair thicker and rougher. Conditioners try to prevent the problem by leaving an oily, water-repelling coating on the hair. Some anti-frizz products accomplish the same thing using silicone, which simultaneously seals the hair and weighs it down, straightening it.

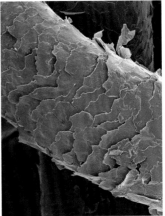

© Photo Quest Ltd/Science Photo Library/Corbis

Your body creates each hair in a *hair follicle*—a tiny pouch deep in your dermis (page 18), where your hair is woven together out of living cells. As each new layer of cells is tacked onto the bottom of the hair, the cells die, and the hair becomes just a little bit longer. In other words, your body treats your hair the same way it treats your skin—it keeps the living cells on the inside and puts the dead stuff on the outside. Which is good in a way, because a head full of living hair would make for an agonizing day at the barbershop.

Note Your hair stores a permanent record of the toxins you ingest, including illegal drugs like cocaine, amphetamines, heroin, and marijuana. A standard hair drug test searches for traces of drugs consumed over the last 90 days. But take longer strands of hair or some slow-growing body hair and you can easily put the last year of your life under the microscope.

Design Quirks

Goose Bumps

Goose bumps are the bumps that appear on your skin (most noticeably on the arms) when you feel cold or experience certain strong emotions, such as fear or awe. The phenomenon occurs when millions of tiny muscles, one at the base of each hair, contract. This makes your fine hairs stand on end.

Humans share this reaction with many other animals, most of whom put it to better use. For example, hedgehogs raise their quills to deter attackers, and cats fluff their fur to intimidate rivals. All animals with fur use this reaction to trap warm air close to their skin. But because our body hair is so short and fine, goose bumps seem to do us no good at all. They're just an evolutionary quirk left over from the days of our hairier ancestors.

Vellus Hair

Compared to other animals, humans appear relatively hairless. But the surprising truth is that you have more hair follicles crammed onto each square inch of your skin than the hairiest chimpanzee, monkey, or gorilla. The difference is that most of your hair (whether you're a man or a woman) is nearly invisible. It consists of a fine, slow-growing, almost colorless covering of downy hair called *vellus hair*.

Vellus hair blankets your body, insulating your skin and heightening your sensitivity to touch. It's the reason you can sometimes "feel" a person moving past you in a darkened room—the passing air currents disturb your fine hairs and trigger the sensitive nerves attached to them. However, vellus hair is easily overlooked and nearly invisible without a magnifying glass. It's sometimes known as "peach fuzz."

Terminal Hair

Terminal hair is the more obvious hair found on your body, including the hair on your head, your eyebrows, and your eyelashes. After puberty, terminal hair appears in many more places on your body—some where it's wanted, and some where it's decidedly inconvenient.

Fun Facts

The Hitchhiker in Your Eyebrow

Your eyebrow and eyelash hair also provide a home for a bizarre, sausage-shaped created called *Demodex*. This creature drags itself along with its eight stubby legs (all at the front of its body) and burrows headfirst into a hair follicle, next to the hair shaft. Apparently, they dislike light, and are much happier anchored in place, eating sebum and dead skin cells.

As far as biologists can tell, Demodex is a harmless traveler. It doesn't spread disease or cause irritation, except in people who have severely compromised immune systems due to some other disease. For unknown reasons, it's very particular about its neighborhood, and is rarely spotted anywhere but on a human face.

Image courtesy of Andrew Syred

Terminal hair is thicker, longer, and darker than vellus hair, although some individuals can have light-colored and fine terminal hair. Terminal hair also boasts a range of textures and colors. The difference between wavy, curly, and straight hair is all in the shape of the follicle that produces it. For example, a perfectly round follicle constructs straight hair, while an oval follicle produces wavy hair. All the hair-care products in the world can't alter the shape of your follicles.

Like most of your body's equipment, terminal hair has good reasons for existing:

- The terminal hair on your eyelashes keeps dirt and insects out of your eyes. Your ear hairs and nose hairs play a similar role.

- The terminal hair on your eyebrows prevents sweat and rain from dripping onto your face.

- On your head, terminal hair helps prevent sunburns on sunny days and heat loss on cold ones.

- Pubic hair is a type of terminal hair that serves as a *secondary sexual characteristic*. That means it's there to advertise that you're a fully functioning adult with the appropriate baby-making abilities.

And, of course, humans have picked up the habit of using hair for something entirely different—as a powerful expression of self-identity that can announce everything from your gender to your political affiliation.

Shampoo and Conditioner

Few products make the bold, imaginative, and highly delusional claims that shampoos and conditioners do. Almost every brand describes mystical powers that can revitalize, energize, volumize, and therapize hair (and the last two aren't even real words).

Unfortunately, the science of hair pours some distinctly unsudsy water on the whole idea. Because each one of your hairs is a sorry strand of dead material, there's really nothing you can do to "nourish" it. That means you don't need to shampoo with vitamins or amino acids. The best botanicals are the ones you grow in pots and water twice a week, and you're better off rubbing herbs and fruit extracts on your dinner than on your scalp.

And forget other hair-care health claims—the government doesn't regulate shampoo, so manufacturers don't need to substantiate their fanciful promises. (For example, some shampoos boast that they protect hair from ultraviolet rays. This typically means that the manufacturer has added a UV-protective ingredient, which you'll only end up rinsing down the drain, and which isn't present in strong enough concentrations to have an effect in the first place.)

The truth of the matter is that shampoo provides a rather straightforward hair-cleaning service. To understand how it works, you need to know how your hair gets dirty in the first place. Ordinarily, the same sebum that lubricates your skin (page 22) moisturizes your hair. This is mostly a good thing, because the thin layer of oil protects your hair from damage. But as the hours pass and you go about your daily business, your hair collects natural oil and skin flakes that shed from your scalp. This is where shampoo comes in—it includes powerful *surfactants* that dissolve these substances, in much the same way that you rinse dirt out of clothes with detergent or grease out of pots with dish soap. The problem is that, in the process, shampoo strips out most of the sebum, leaving your hair dry and fragile (although the effect is far gentler than if you showered with laundry detergent or dish soap).

To balance this effect, many shampoos have conditioning agents, and many people use a separate conditioning product. There's a bit more variability to the way that conditioners work, but essentially they all aim to coat the hair shaft with protective sebum-like compounds. Some creamy conditioners feel heavy in the hair and glue together damaged fibers and loose scales. Other conditioners are lighter and oilier. But all these substances cling to the hair shaft and don't rinse out with plain water.

The best hair-care advice for a biology wonk is this: Don't break your budget on high-end products. Buy the shampoo that matches your hair type (oily or dry) and use conditioner to manage excessive dryness. Finally, don't pressure yourself into washing your hair every day. If you're just as happy waiting a day or two, your dead hair will probably be a bit better off.

Hair Growth and Hair Loss

Your hair follicles have a tiring job, and every once in a while they take a break. At the moment, roughly 90 percent of the hair on your head is growing, while the rest is taking some time off. Some of that hair will resume growing again after a pause of a week or two. A smaller proportion will simply fall out—about 50 hairs a day. But don't panic, because the same hair follicle will begin creating a new hair in its place. An average hair takes six years of abuse on your head before it drops out and the hair follicle starts over.

Eyebrows and eyelashes have a different growing schedule. Eyebrow hairs grow for about 10 weeks, and then rest for the better part of a year. (This is what makes eyebrow shaving such a dastardly revenge tactic.) Eyelash hairs last about three months apiece before falling out and being replaced.

> **Note** Hair growth is an issue that comes with a boatload of baggage. Hair embarrasses us when it appears in certain places (inside our ears, for example). In other places, it mortifies us when it vanishes. But other than cutting your hair, you have little control over its comings and goings.

Here are some quick facts that can help separate the bare facts from the follicle folklore:

- Hair doesn't grow faster or thicker after you shave it (on any part of your body).

- Hair doesn't grow faster at night. Female hair doesn't grow faster during menstruation. Instead, all hair grows at a constant rate with a brief resting period.

- Frequent washing, blow drying, and dyeing your hair doesn't destroy hair follicles or slow hair growth. However, these activities might make your current hair more brittle and fragile. But even if you damage a hair to the point of falling out, the same hair follicle will produce a new one to take its place.

- You're born with all the hair follicles you'll ever have. As you grow and your skin stretches from infant-sized to adult proportions, your hair follicles simply become more spread out.

- While you're pregnant, each hair clings on a little bit longer, eventually giving you a fuller head of hair. After you give birth, your body sheds its hair more quickly to make up for lost time.

- The only ways to remove hair permanently are laser hair removal and electrolysis. Both treatments take numerous sessions over the course of many months, and neither treatment works for all people or all hair.

- Wearing hats doesn't cause hair loss.

Male-pattern baldness, which causes the infamous ring-around-the-bald-spot effect, develops gradually and eventually affects about two-thirds of all men. Its causes are genetic, and its treatments are few. A small set of medications give some improvement to some people, but these drugs are often ineffectual. There are only two guaranteed solutions: hair-transplant operations (which are expensive, time-consuming, and may look odd, since hair loss continues around the transplanted patches), and head shaving. If you opt for the latter, you'll likely tell people that you deliberately chose baldness to emphasize your virile, youthful manliness. Everyone will know the truth, of course, but they'll also be quietly relieved that you aren't practicing the dreaded comb-over.

The Psychology of Comb-Overs

The *comb-over* is a hair-grooming practice in which a balding man brushes a few strands of hair over a wide expanse of his bald head, usually starting with an unnatural part. Sometimes he cements the hair in place with oil or a styling product. The hallmark of a comb-over is that the combed-over hair covers only a small portion of the available scalp area. Much as a piece of avant-garde music might call attention to the silences between successive notes, a comb-over directs your helpless attention to the hair that is no longer there.

Comb-overs are a somewhat mysterious phenomenon. Although most men find them distasteful, many still end up adopting them in later years of baldness. Sociological thinkers (and people with a great deal of extra time on their hands) suggest that comb-over practitioners fall prey to the *sorites paradox*. Essentially, the sorites paradox describes how small steps that seem sensible on their own can lead to an absurd outcome. In the case of comb-overs, the victim may begin moving the part of his hair by a small amount to add fullness to a region of thinning hair. Only as the process of baldness accelerates does this become a futile attempt to hide a glaring patch of skin under the last few stragglers of hair.

Incidentally, the Japanese call men with comb-overs *barcode men*, because the lines of neatly aligned hair resemble barcode symbols.

2 Fat

I t can strike fear in the heart of the most level-headed, body-positive person. Wrapping your body just under its outer covering of skin is a gentle, gelatinous blanket of *fat*. Serving as insulation, cushioning, an energy reserve, and the focus of intense social scrutiny, fat is one body component that the average person spends more effort to remove than to understand. But fat is no lightweight—although it gets a lot of bad press, it's as essential to your survival as any of your more popular organs.

Fat is also at the heart of a controversial body mystery. Unlike the fine-tuned processes in the rest of your body, fat storage is the one mechanism that frequently goes completely off the rails. In the process, excessive fat sets up ordinary people for a dismal collection of health troubles.

It's hard to overstate just how big the problem is. Compared with other animals, obese humans are biological wonders—pound for pound, the fattest creatures on earth. (If you aren't already feeling self-conscious, consider this: The percentage of body fat of the fattest humans tops that of even the generously proportioned beluga whale.) Still more remarkable is just how *common* excessive fat is. Despite billions of dollars, high-powered research, and some seriously good intentions, people are getting fatter year after year, in countries across the globe. In the U.S., more than a third of the population is overweight and another third is obese, leaving less than a third of the population to give everyone else disapproving looks in line at the all-you-can-eat restaurants. Clearly there's something about the modern world that's throwing the body's carefully tuned mechanisms seriously out of whack.

In this chapter, you'll start to unravel this riddle. You'll learn why you need fat, how it works, and why your body is so keen to hold on to every ounce of fat it's got. You'll also measure your body-fat percentage and learn some techniques that can help you battle excess weight. Along the way, you'll explore some common questions, like: Do diets really work? Who's to blame for your bulging belly? And why do some people struggle to count calories while others can down an entire cheesecake and get away scot-free?

The Purpose of Fat

Fat elicits contradictory emotions. Oh, we like it well enough when it coats our fish sticks, but it's not nearly as appealing when it's wrapped around our midriffs. And even though fat's a visceral part of us, we treat it as an unwelcome intruder. Consider how most well-adjusted people can agree with both of these statements:

> Every part of the human body is good, proper, and has its own beautiful purpose.

> Fat is an evil, evil thing, and I'd do anything to eliminate it.

But before you call the local liposuction clinic, consider this: Fat—a modest amount of it, anyway—is your friend.

The fat in your body cushions your organs and joints, absorbing shocks and protecting them from damage. It also insulates your body against temperature extremes. Less critically, body fat gives you a cushion to sit on and covers hard bone with padding, making hugs much more pleasant. The fat in your diet helps your body absorb certain vitamins, produces compounds that regulate blood clotting and inflammation, and builds structures like cell membranes. Your brain is literally filled with the stuff—by weight, your brain's more than half fat.

Most important, fat is the body's *energy storage system*. In times of plenty, your body stores extra calories as fat, ensuring that none of the effort you spend cooking, chewing, and digesting goes to waste. When food is scarce, your body taps into its fat reserves, burning them to fuel everything your cells do.

This energy storage system kept our distant ancestors alive through tough times and countless famines, while the nibblers and picky eaters perished. The problem is that famines don't come along as often as they used to. As a result, average people spend most of their lives planted firmly in the first part of the energy-storage equation, hoarding fat for food shortages that never come.

Big Fat Myths

Fat is always on our minds. We waste many productive hours thinking about the fat that's on our plate, our bodies, other people's bodies, and (for purely scientific reasons) the bodies of scantily clad celebrities on the Internet. Other body parts don't come close to getting this much attention, which is probably why fat is the subject of a messy collection of mistakes, myths, and misinformation. Here are some fat factoids you might have encountered:

- **Muscle turns to fat.** Muscle and fat are different types of tissue, and the body can't convert one to the other any more than you can transform a pear into a chocolate truffle. However, your fat sits on top of your muscles, which means that a small weight gain can quickly mask those six-pack abs.

- **A calorie is a calorie.** There are a dizzying range of factors that can influence how well your body transforms food into energy and (ultimately) fat, from the quality of the bacteria in your digestive tract (page 196) to the combination of nutrients in the food you eat. So-called experts that miss this subtlety warn that a single extra glass of juice a day can add up to dozens of pounds of weight gain over a year, which is clearly nonsense. Other than paranoid dieters, no one eats with such maniacal precision. The truth is that the body is miraculously successful at preserving its preferred weight, as explained on page 55.

- **Thin people have faster metabolisms.** This is partly true, in the sense that naturally thin people often expend more energy in small, natural ways (for example, with constant fidgeting). However, people of all body types appear to burn calories in roughly the same proportion to their weight, which means that obese people actually need *faster* metabolisms to maintain their extra poundage. However, there's one huge exception: If you start an extreme diet, your metabolism will fight you, slowing itself down to a calorie-conserving crawl. Similarly, pig out in an effort to put on pounds, and your metabolism will speed up to keep you at your current weight. Page 55 has the details.

Fat Up Close

Your body has billions of fat cells. (For example, people with healthy, normal weight carry around some *40 billion* fat cells.) Each one is a miniature reservoir for storing fat.

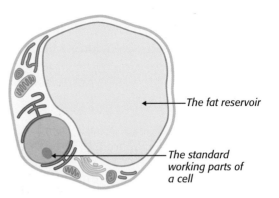

The fat reservoir

The standard working parts of a cell

Fat cells work in a rather unusual way. When your body squirrels away fat, your fat cells don't do what you might expect them to—multiply. Instead, they gobble up the extra fat, inflating themselves like rubber balloons. In fact, if you take a close-up look at a fat cell under a microscope, you find that nearly the entire cell is filled with a greasy droplet of fat. The rest of the cell—the cellular machinery that runs the shop—is squeezed to the very edge of the cell, and much more difficult to spot.

> **Note** Fat cells make up a spongy type of tissue that biologists call *adipose tissue*. Inside this tissue, your body packs fat cells together in a neat, almost honeycombed pattern. It's like bubble wrap for your body.

Early researchers wondered if fat cells actually did any work on their own, or if their squashed functional parts were completely disabled. As you'll discover in the next section, your body fat is busier than you think.

How Fat Controls Your Body

In the past, scientists thought fat cells were simply boring blobs of lard. We now know that they don't hang around idly—instead, they release massive quantities of hormones that have effects throughout your body. For this reason, many fat scientists (er, make that scientists *studying* fat) argue that fat isn't just excess tissue—it's a smart and powerful organ in its own right.

Note *Hormones* are special chemicals the body uses to send messages from one place to another. Your fat releases hormones into your blood, where it can travel to other body parts, like your brain.

At this point, it's natural to ask what these fat-secreted hormones are up to. Although the interaction between fat cells and the rest of your body is fantastically complex, here are some key examples:

- **Female fertility.** As women know, carrying and delivering a baby is a body-straining odyssey. Without a bare minimum of fat, your body won't even let you try. If your body decides that you don't have the necessary fat reserves to live through a pregnancy and nourish a baby, you'll stop menstruating. Regain the fat, and the deal's back on.

- **Appetite.** Fat cells release a hormone called *leptin* that tells your brain to damp down feelings of hunger. Leptin is often blamed for the nearly inevitable weight gain that follows severe dieting. As you lose fat, your leptin level falls and your brain feels less satisfied, becoming much more likely to trigger a late-night cheesecake craving.

Note It might seem that a little extra leptin would make a great diet pill. Unfortunately, it doesn't work that way. The problem is that your body is far more interested in preventing starvation than in fighting weight gain. If your leptin level rises, your brain adapts to this new level and considers it the "new normal." Then, when you stop taking the pills and your leptin level falls, hunger quickly sets in. This process, like much of the body's appetite-control system, is a bit biased. In fact, it looks a lot like a one-way street to more eating.

- **Regulating the immune system.** Fat cells release compounds that fire up parts of your immune system. Too much fat, and these signals get amplified to harmful levels, triggering *inflammation* deep inside your body. This inflammation can damage your body and lead to other conditions, like heart disease, arthritis, and type 2 diabetes. On the other hand, elite athletes who eat an extremely low-fat diet—say, ultra-marathon runners—end up *suppressing* their immune systems and becoming more susceptible to infection.

The Secret Power of Fat

The connection between fat and hormones is at the forefront of fat research. In the past, scientists assumed that the problems caused by excess fat were the result of its physical burden. They worried that fat would strain bones and joints, plug up arteries, and crush internal organs. But today, many scientists believe that the chemical contribution of fat is more important—and far more dangerous. Although the hormones that fat secretes are critically important in small doses, in large amounts they can trigger runaway inflammation, confuse other organs, and throw off your body's regulatory processes.

Here's an example: *Osteoarthritis* is a painful condition that causes the cartilage in your joints to break down. Studies find that as your body weight creeps up, your risk for osteoarthritis rises along with it. The traditional explanation is that excess weight causes extra wear and tear to important joints like your knees, which is certainly possible. However, increased body weight also increases the likelihood of osteoarthritis in places where weight isn't all that significant, like your hands. A more recent explanation is that excess fat triggers chemical changes that can wreak havoc on your body (for example, inflammation that attacks the cartilage in your hands).

The Life and Death of a Fat Cell

As you've seen, fat cells are like miniature people—the more fat you feed them, the more they plump up. Unfortunately, even the strictest diet can't remove them. Instead, it leaves you with billions of slimmed-down fat cells crying out for another meal.

But there's another side to this story. Different people *do* have different numbers of fat cells. Not only do obese people have dramatically bigger fat cells than slim people, they often have many more. For years, scientists have wondered what causes these differences. Do fat cells ever die? Does something trigger the creation of new fat cells? Is it our fault?

In 2008, a group of scientists devised a clever experiment to answer these questions. It worked by examining the effect of nuclear-bomb explosions on human fat cells. At first glance, this sounds like a Very Bad Idea. After all, detonating nuclear bombs— even in the name of science—is likely to put off the ethics board. But here's the trick: the researchers didn't launch the bombs themselves. Rather, they took advantage of a bit of Cold War history to track *radioactive carbon* in human fat cells.

CAUTION

RADIOACTIVE
MATERIAL

Now you're probably wondering what radioactive carbon is doing in human fat cells, and that's a reasonable question. The obvious answer is "because the researchers put it there," but it turns out they aren't allowed to do that, either. (It's likely to be toxic.) The experiment was launched, unknowingly, by the American and Soviet militaries when they tested nuclear bombs. These tests sent large amounts of radioactive carbon into the atmosphere that drifted to all corners of the world. Plants pulled it out of the atmosphere, animals ate the plants, and we ate the animals. The end result is that the U.S. and Soviet governments radioactively labeled pretty much everyone who was around at the time.

Here's a chart that shows how the levels of radioactive carbon soared during nuclear testing and rapidly diminished when nuclear testing went underground in 1963.

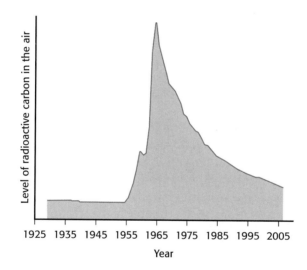

By matching the level of radioactive carbon in the air with the amount of it in the DNA of a fat cell (along with a generous pinch of statistical mojo), researchers can figure out when that cell was created. It's as though each fat cell has its own "manufactured on" date.

To draw their conclusions, the researchers analyzed fat extracted from about 700 people. Here's what they found:

- The number of fat cells grows through childhood and adolescence, but stabilizes sometime in late adolescence. After that, your fat-cell count stays the same.

- People on the fast track to obesity pack on their fat cells more quickly in childhood and stop producing them around age 16 or 17. Naturally thin people have fewer fat cells, but keep producing them until the age of 18 or 19.

- Even extreme events—for example, dramatic weight loss through surgery or super-sized weight gain—change the size of fat cells in the body, but don't nudge the number. (That said, many experts believe years of excessive weight gain that lead to *hundreds* of extra pounds will eventually cause fat cells to multiply.)

- About 10 percent of your fat cells die and are replaced each year, whether you're thin, fat, or somewhere in between.

No one knows why some people end adolescence with more fat cells than others. There's a good case to be made that it's hard-coded in your genes (in other words, blame mom and dad). Events during fetal development and childhood might also play a role—for example, chowing down on calorie-rich food early in life could kick off certain processes in your body, preparing it for a life of fat retention. But no matter the reason, it's clear that some people enter adulthood set up for caloric challenges, with a big family of fat cells. They won't necessarily gain weight more easily than a naturally lean person, but they'll probably feel a stronger pull toward that second batch of chocolate chip cookies.

To put it another way: fat people *are* doing something wrong. But they're doing it because their bodies are telling them to.

The Lessons of Fat Cells

Finding out the secret rules of fat cells is fascinating, but does it do you any good? The obvious hope (both for failed dieters and pharmaceutical executives) is that this understanding will lead to a drug that can control appetite or fat storage. In the meantime, you can learn a few things from your fat cells:

- **Hold the line on childhood eating.** No one knows why different people gain different numbers of fat cells throughout childhood and adolescence. However, there's at least a possibility that environmental influences are at work, meaning early binging habits might tune the body to a lifetime of calorie craving. So if you're a parent, make every effort to provide a varied, healthy diet for your child—one that's rich in fruits and vegetables, and low in processed foods—and resist the temptation to teach "clean your plate" or "food is your reward" lessons. Even if your efforts don't influence your child's fat cell count, they'll help set down lifetime habits that can defend against the worst dietary excesses.

- **Expect diets to be difficult.** Because a diet can't change your fat-cell count, you'll always have the potential to regain the weight you lose. To give yourself the best odds of staying slim, start by dieting small (with a goal of losing 10 pounds at a time), then concentrate on maintaining your weight and adjusting to a new lifestyle.

- **Don't moralize fat.** While it's a biological fact that fat people get fat by eating too much, none of us is completely in control of the powerful drive to eat. Those with a larger collection of fat cells just might start life with the odds stacked against them.

Taking Stock of Your Body Fat

If you've ever struggled with obesity, or even mild pudginess, you've probably spent a lot of time thinking about a single statistic: your weight. After all, there's no easier way to size up just how much of you there is than to step on a scale.

Fortunately, scales don't have a memory or offer feedback. (One can imagine what they might say: "Oh my. Gained another five this week? You'd better make sure your wife doesn't catch wind of this.") As a result, it's up to you to decide whether the three digits on the scale add up to good news, bad news, or Seriously Bad News. In the following sections, you'll learn about several measurements that can help you reach a verdict.

Body Mass Index

One common, but somewhat crude, measure for assessing your poundage is the *body mass index*, or BMI. Oddly enough, the BMI was created by a Belgian mathematician (perhaps concerned about overindulging on the chocolate delicacies of his countrymen). In any case, the BMI became a surprisingly popular and somewhat controversial way to separate the thin from the fat.

To calculate your BMI, you divide your height (in meters) by your weight (in kilograms). Or, you can plug more common pound and inch measurements into this version of the formula:

$$BMI = \frac{weight\ (\text{lb}) \cdot 703}{height^2\ (\text{in}^2)}$$

This gives you a single number that ranks your weight. For example, if you're five feet eight inches tall (that's 68 inches total) and weigh 180 pounds, your BMI is 27.4. According to the standard BMI groupings, that puts you into the overweight category.

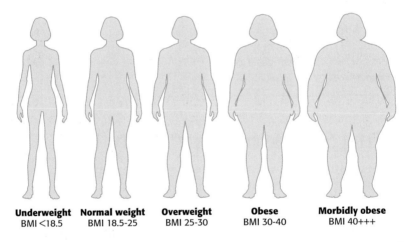

Underweight	Normal weight	Overweight	Obese	Morbidly obese
BMI <18.5	BMI 18.5-25	BMI 25-30	BMI 30-40	BMI 40+++

If you don't want to pull out your pocket calculator, you can use one of the many BMI calculators online (just search for "BMI"). Or you can find your spot on the handy BMI chart shown on the next page, which helps you judge exactly where you fall in your weight class.

The BMI has some known weak spots. For instance, short, muscular types and athletes can end up in the obese zone even though they're in prime condition. Similarly, elderly folks can coast through with a normal ranking if they have high body fat combined with very little muscle weight.

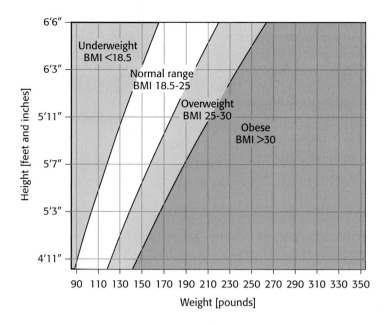

Despite these weaknesses, the BMI is good for two things:

- **Making conclusions about a population.** For example, if the BMI suggests that one-third of Americans are dangerously obese (as it does), the odds are that very few of them are muscular athletes in the prime of their lives.

- **Giving a rough idea of the weight situation for an ordinary person.** If you fall far outside the normal zone, the BMI is giving you a red warning flag. It's up to you to follow up and see if you really do have the weight problem it suggests. The standard next step is to measure your waist (page 53) and analyze your blood (see page 49).

Body-Fat Percentage

If you're concerned about good health, your sheer poundage isn't nearly as important as finding out how much of you is made up of fat. In other words, you need to spend less time thinking about your weight and BMI, and more time concentrating on your *body-fat percentage*—the weight of your fat compared to the total weight of your body.

Everyone has a certain amount of essential fat stored in small amounts in organs, bone marrow, muscles, and the nervous system. This fat supports the normal functioning of these systems. Women have a bit more fat in the breast, pelvis, hips, and thighs, which is a prerequisite to making babies. Along with this bare minimum, it's important to have at least a little more fat to use as an energy reserve, so you don't collapse the next time you skip breakfast.

Description	Women	Men
Essential Fat	12-15%	2-5%
Athlete	16-20%	6-13%
Ordinary Fit Person	21-24%	14-17%
Acceptable	25-31%	18-25%
Obese	Over 32%	Over 25%

Unfortunately, the only place you can get a foolproof measure of your body-fat percentage is on an autopsy table. Some other techniques are nearly as good, but require the work of professionals and expensive hospital equipment. A few are much less accurate, but can be carried out at home. Here's a quick roundup of the ways to measure your body fat:

- **Imaging.** An MRI (magnetic resonance imaging) machine can peer under your skin to create shocking scans that show the amount and distribution of your body fat, like the one on page 53.

- **Hydrostatic weighing.** Some specialized laboratories (and a few health clubs) have water tanks that are designed for underwater weighing. This technique works because fat isn't nearly as dense as muscle and bone (which is why well-padded people float more easily). Using a bit of math, you can combine your underwater weight with your normal weight to get a fairly accurate measurement of your body-fat percentage.

- **Skinfold measurements.** To perform this test, a professional painstakingly pinches the folds of fat in various places on your body using calipers. It's not particularly accurate, and it doesn't sound like anybody's idea of a fun Friday night. However, it works well if you just want to monitor changes in your body fat, as long as you get the same person to administer the measurement each time.

- **Electrical conductance.** This test sends a faint current through your body, calculates your body's electrical resistance, and uses the result to estimate your body-fat percentage. This works because muscle contains a lot of water, and so conducts electricity quite well, while fat does not. This technique is used in hospital-grade equipment that costs tens of thousands of dollars, and in cheapie department-store versions, which look like ordinary digital scales.

Do-It-Yourself Fat Measurement

A body-fat scale (which looks like your average bathroom scale, but measures electric conductance instead of weight) is the most affordable, practical way for you to keep an eye on your fat without getting help from someone else. But even top-rated scales make so many assumptions that you can't trust the number they give you. So does that make them a gargantuan waste of time?

Not necessarily—provided you use your body-fat scale to measure *changes* in body fat. Used this way, a body-fat scale lets you judge the success of an exercise plan or the toll of an expanding waistline. To use this strategy, you need to make sure you get a consistent reading. Here are some tips that can help:

- **Watch your water intake.** The amount of water in your body can dramatically change the reading of a body-fat scale. The best idea is to measure your body fat at the same time of day, at least one hour after drinking or eating, and definitely not after exercise. Consider making it part of your morning routine.

- **Adjust the settings.** For example, many scales have a specific profile for athletes or children. If you pick the right settings, the scale is more likely to make the right assumptions for your body and give you a more accurate reading.

- **Don't compare people.** You and your friend may use the same scale, but it's not fair to compare the numbers. Minor differences between the two of you can skew the results in different ways. Similarly, it doesn't make sense to compare your own readings on different body-fat scales.

- **Get a baseline.** If you get the chance, measure your body fat with a more reliable method—for example, a skinfold test or an underwater weighing—then compare that result with the reading on your scale to figure out a basic frame of reference for how accurate your scale is.

Bring Your Fat to a Checkup

Your weight, BMI, and body-fat percentage provide some important clues about your overall health, but they're far from definitive. For more comprehensive information, it's time to doctor up.

Yes, you may hate waiting in the doctor's office. And you probably aren't crazy about stripping off your clothes and donning a paper gown that's roughly the width of two paper towels. But your family doctor is an important team player in every aspect of your body's health. If your weight has been creeping up and you're starting to get nervous about the possible ill effects, your doctor can run some of the following tests:

- **Cholesterol levels.** Doctors measure two types of cholesterol. At high levels, *LDL cholesterol* can clog your arteries when it forms into a sticky plaque. *HDL cholesterol* is the good stuff—it helps mop up excess LDL cholesterol, keeping your arteries clean. To stay healthy, you need a balance between these two players. A blood test can tell you how your body is doing.

- **Triglyceride level.** *Triglycerides* are a form of fat that travels through your bloodstream. Your body uses triglycerides to move fat from one spot to another, so it can fuel the work going on in your body. But if there's too much fat circulating in your blood, you're at greater risk for heart disease. Once again, you can check your triglycerides with a blood test.

- **High blood glucose.** Normally, the human body is extremely effective at pulling excess sugar out of your blood using *insulin*. But if this sugar-storage system is starting to malfunction—perhaps after a lifetime on the Krispy Kreme diet—your blood will stay sugared-up. This is the beginning of diabetes. Left unchecked, the runaway sugar can damage your heart, eyes, and kidneys.

- **Blood pressure.** Blood pressure measures the force of your blood as it pushes through your arteries (page 161). Although high blood pressure causes no immediate damage (and has virtually no obvious symptoms), over time it strains your heart and thickens your arteries. Left untreated, it can lead to organ damage, heart attack, or stroke.

Even if you have an ideal BMI and aren't worrying about your weight, there are reasons you might still want to take these tests. For example, you might have a family history of early heart disease, smoke two fistfuls of cigarettes between breakfast and lunch, or have unexplained symptoms. If in doubt, chat it up at your next physical.

Where Fat Lives

Your body has plenty of hiding places to stash fat. Where fat goes is completely beyond your control—it depends on your gender, age, and genetic inheritance. But here's the unfair part: even though you can't choose where your body stores its fat, those locations can have serious consequences for your health. Studies show that the risks of excess fat increase dramatically when it's stored in certain places.

Essentially, there are two basic storage zones for your fat:

- **Subcutaneous fat.** This fat is stored in a layer around your body, just under your skin. Once again, genetics determine where the padding is thickest. In women, common storage depots include the thighs and posterior. In men, fat is more likely to hang out around the abdomen. Subcutaneous fat is also responsible for cute baby faces and the dreaded *cellulite* (lumpy deposits of fat that cause a dimpled appearance in skin, usually in women).

- **Visceral fat.** This fat is buried under your muscles, deep inside your body. It pads the space around your internal organs. Most experts believe visceral fat is far more dangerous than subcutaneous fat, and leads to a higher risk of heart disease and diabetes.

Subcutaneous fat is by far the more popular fat parking zone, accounting for the majority of fat in anyone's body. Smaller quantities of fat can also end up stored in your muscles (where it's called *intramuscular fat*) and inside certain organs (such as the liver), where it's especially dangerous.

There's no consensus on why visceral fat is the fat to fear. It's possible that the hormones it releases interfere with the functioning of nearby organs or compels them to start storing their own fat reserves, which can cause further damage.

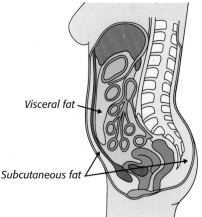

Visceral fat

Subcutaneous fat

Note No matter what foods you eat or exercise you practice, you can't remove fat from a specific part of your body. In fact, studies have shown again and again that spot reduction is impossible. Lifting weights will not cut the flab from your arms. Doing stomach crunches will not shrink your belly. And so on. Toning your muscles may improve the appearance of an otherwise chubby part of your body, but that's all you can hope for. Incidentally, when you diet, the last place you gained fat is the first place you'll lose it.

Body Shapes

You probably already have an inkling about where your body stores its fat. But if you're overweight, it's important to know how much of your excess padding consists of hazardous visceral fat. To get the final word, you'd need to scan an image of your body with a high-priced MRI machine. But even if you don't have any hospital equipment in your basement, you can still look for a few clues, such as your belly width and body shape:

- **Apple-shaped bodies** have more fat around the abdomen and the chest.

- **Pear-shaped bodies** hold a lot of their extra fat on the outside—primarily around the thighs and bottom.

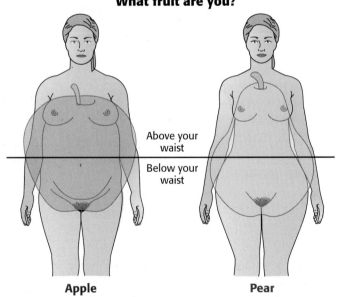

What fruit are you?

Above your waist

Below your waist

Apple   Pear

Although apple-shaped bodies have their share of subcutaneous fat, they're more likely to store dangerous quantities of visceral fat. This makes sense if you remember where your body stores visceral fat—in the space around your abdominal cavity. As visceral fat plumps up, it pushes the rest of your body out, creating the distinctive apple shape. The following MRI scan reveals the inner world of fat.

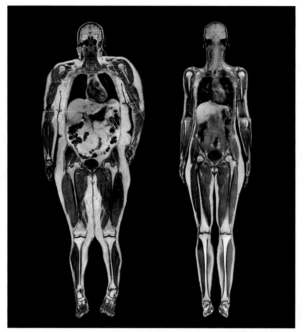

Marty Chobot/National Geographic Image Collection

Note Everyone has subcutaneous *and* visceral fat, no matter what their body type. Although the apple body shape is a warning flag, pears aren't necessarily safe. Another worthwhile check for visceral fat is to measure your waist. If you're a woman with a waist that measures 35 inches or more, or a man with a waist that's 40 inches or more, you have a significantly higher risk for diabetes and heart disease.

Fighting Visceral Fat

Now you know about visceral fat—the greasy matter that's buried deep inside your abdominal cavity, packed around your internal organs, and up to no good at all. The obvious question is this: What can you do to get rid of this accursed stuff?

Although the science isn't settled, several studies suggest that dieting might be less helpful than you think. The problem is that it's likely to shrink your subcutaneous fat without reducing the dreaded deposits of visceral fat that lurk inside. Instead, you need to add high-intensity exercise to the mix (say, 30 minutes of pulse-racing activity, four times a week) to pare down visceral fat. Some experts believe that even happily thin people can have hidden deposits of visceral fat (see the box on this page) and need the same exercise regimen to stay healthy.

Tip Think health gain, not weight loss. Most people find it easier to concentrate on doing something good than on trying to stop something that's gone wrong—especially if you shoulder hefty feelings of guilt for the pounds you've already put on. So make fitness your goal, and use weight loss and exercise as just two of your tactics to get healthy.

Design Quirks

Thin on the Outside, Fat on the Inside?

It's every string bean's worst nightmare: What if natural, effortless skinniness isn't the hallmark of health, but a socially acceptable veneer hiding some of the same health problems?

If this idea troubles you, thank Dr. Jimmy Bell, a British scientist who has scanned more than 800 people with MRI machines to study their fat (creating pictures like the one shown on page 53). One of his most disquieting discoveries was that nearly half of people with normal-weight BMI scores actually had excessive amounts of visceral fat buried inside their bodies. Apparently, those most at risk for hidden visceral fat are people who seldom exercise and maintain their weight through diet alone. As Dr. Bell puts it, "Being thin doesn't automatically mean you're not fat."

While this should worry the skinny, it isn't much comfort to the obese. They're even more likely to have still larger deposits of visceral fat. The only exception is if they're extremely athletic. For example, some studies suggest that sumo wrestlers, despite their proudly displayed subcutaneous fat, actually have a fair bit of muscle and surprisingly little visceral fat.

Dr. Bell's discoveries are controversial, and they call into question virtually everything we say about fat. After all, if you can have health-damaging amounts of fat without *being* fat, how can anyone really get a handle on the danger of these dastardly cells?

There's one practice that definitely won't help: *liposuction*. Although getting the fat vacuumed out of you seems like a fiendishly convenient shortcut, liposuction sucks up subcutaneous fat only. This is probably why it has no long-term health benefit. Studies show that happily liposuctioned people don't do any better on key measures of health and inflammation (including the blood tests listed on page 49) than others. This is true even when massive quantities of fat are removed—in one study, 20 pounds at a time.

So far, there's no way to liposuction away your visceral fat. However, small experimental studies have found ways to surgically remove it. In one study, visceral fat was removed from a group of obese men with insulin resistance, the precursor to diabetes. Even though they stayed obese, with copious amounts of subcutaneous fat, almost all of them lost their insulin resistance within a year.

Understanding Your Body's Anti-Starvation System

You've already learned how fat cells use leptin (page 41) to talk to your brain, making it difficult to reduce your waistline without increasing your appetite. If that were the entire story, the science of food and dieting would be a tidy affair.

Instead, the processes that regulate eating aren't fully understood. They're also multilayered, which means that your body uses multiple, redundant processes to get you to the dinner table. If something throws off the leptin signals, another system fills in to make sure you still feel hungry and get a good meal. This is one of the reasons that there's no miracle pill for obesity on the way. When scientists attempt to tweak one set of hormones to promote light eating, the body adapts and chows down as steadily as ever.

Naturally Lean and Naturally Fat

The discovery that different people harbor different numbers of fat cells hints at one of the unalterable truths about obesity. Some people struggle with weight their entire lives, while others wonder what all the fuss is about.

Truthfully, this shouldn't come as a surprise. A long list of fascinating studies illuminates the strange workings of human fat storage. Here are some of the best:

- **Starving the objectors.** During World War II, more than a hundred men volunteered for a starvation study to avoid military service. Ancel Keys, a health researcher, took the 36 who were the most physically fit and psychologically healthy, cut their food intake to less than half, and put them on an intense walking regimen—similar to the punishing ordeals of the most hard-core dieter. The men lost a quarter of their weight over six months, and then were allowed to gain it back. During the dieting phase, they became obsessed with food, fantasizing about it for hours, collecting cooking implements, and scouring garbage cans. A few men broke the rules with sudden binging, after which they felt nauseated, depressed, and disgusted with themselves. And when they were allowed to choose their own meals again, many were no longer satisfied with normal portions, and ate insatiably. All of these details sound like a natural response to extreme circumstances—but they also have an uncanny similarity to the experiences of crash dieters.

- **Feeding the prisoners.** Another study took the opposite approach. Ethan Sims started with normal-weight volunteers from a state prison, and deliberately overfed and underexercised them until they grew fat. Much to everyone's surprise, fattening people up was nearly as challenging as slimming them down. Different prisoners needed different amounts of excess calories to gain weight, but most of them required huge amounts of food to gain a significant amount of extra weight. At the end of the study, the prisoners effortlessly returned to their original weights—again at varying, self-determined rates.

- **Comparing twins.** It's no surprise that people with obese parents are more likely to be obese themselves. There are a range of possible explanations, including poor dietary habits and bad role models. But more intriguing studies examine adopted children and identical twins who were raised in separate families. These studies found that fatness has a powerful genetic link—in other words, children are likely to drift toward the body mass of their birth parents, even if they've never met them.

Note For popular accounts of fat research, check out these two slim books: *The Hungry Gene* (Ellen Ruppel Shell) explores the science and business of obesity research. *Rethinking Thin* (Gina Kolata) covers some of the same ground, while following a group of dieters in a weight-loss study who have big dreams but are destined to come up short.

Faced with this evidence, it's natural to wonder what some people have done to deserve the short (and fat) end of the genetic stick. The often-repeated line that skinny people have fast metabolisms doesn't seem to be true. Instead, people of all body weights maintain the same weight through most of their lives, eating an appropriate amount for their size. However, when people are pushed out of that range with forced starvation or overfeeding, the body fights back, making meaningful, long-term weight change extraordinarily difficult. This doesn't mean that a fat person can't get thin, but it certainly changes the rules of the game.

How to Supersize a Human

If all this sounds colossally unfair, well, it is. But even if the genetic deck is stacked against you, you'd probably escape unharmed were it not for the cooperation of a seriously skewed environment.

After all, obesity is a recent phenomenon—so recent that worldwide obesity statistics stretch back less than a generation. In the wild, virtually no other species experiences true obesity. Not only do humans break the mold, we also have the dubious distinction of being the only species to bring obesity to other animals, such as our overfed pets, livestock, zoo dwellers, and experimental subjects.

So what's gone wrong in our world to shift a huge number of people—the majority in many countries—into excessive weight? Although there's no definitive answer, it seems overwhelmingly likely that a set of factors work together to help make us fat. The following sections explain what they are.

Changes in activity

For tens of thousands of years, humans spent the better part of the day in activities we'd call exercise, with only the occasional period of rest. Today, the balance has neatly flipped, and our bodies are left sitting around like an idling automobile engine for most of the day.

Changes in food

At the same time that we've switched from lives of restless activity to permanent relaxation, we've changed the way we fuel our bodies. Generations ago, we had no choice but to fill up with natural, unrefined foods that took a fair bit of digestive effort to convert to energy (and fat). By comparison, even the healthy eaters of today down huge quantities of processed, high-calorie, high-carbohydrate foods, which contain enough energy for an Olympic rowing team.

The breakdown of food culture

Studies show that people who eat a diet traditional to their native culture fare better than average eaters. This is true whether the culinary tradition is Greek, Italian, Japanese, French, or something else—even though all these diets are dramatically different and many emphasize sometimes stigmatized foods like pasta, rice, fatty meat, or butter.

There are two possible explanations. One is that the diet most Westerners eat—heavy on processed convenience foods—is, biologically speaking, the worst possible rubbish we could use to fuel ourselves. The other possibility is that people who are part of a traditional food culture are also guided by firm, unwritten rules that govern acceptable eating practices. For example, consider the so-called *French paradox*—the fact that French people have a relatively low incidence of heart disease, despite enjoying a diet rich in cheese, cream, and other sources of saturated fat. Unlike North Americans, French eaters are part of a dining culture that emphasizes slow, shared meals. It discourages second helpings and snacking.

And if you think you've got it bad, consider the plight of cultures that have moved from ancient cuisine to Western convenience foods in a single generation. Two examples are the desert-dwelling Pima Indians and the settlers of the tiny Micronesian island of Kosrae. Both populations are small, genetically similar, and cut off from the rest of the world. For generations, they lived through periods of intermittent famine that were more severe than those faced by most Europeans. Today, they've adopted a Western diet and suffer from staggering rates of obesity and diabetes, far worse than the rest of modern society. Scientists are still battling over whether the problem is environmental (for example, the population is particularly susceptible to Western conveniences because their old way of living and eating is obsolete) or genetic (perhaps they have a higher incidence of genes that spur overeating).

Environmental programming

It's a bit speculative, but many obesity researchers believe environmental cues during pregnancy or childhood activate certain genes in the body, putting a process in motion that ultimately leads to adult obesity. It's not a far stretch—after all, not only have people been fattening up generation after generation, they've also been growing much taller and reaching sexual maturity far sooner. In studies of strains of obese rats, a brief period of underfeeding in infancy reduces their fatness in adulthood. A similar diet in later life has only temporary effects.

Design Quirks

Obesity: The New Kid on the Block

Most people are so familiar with the dangers of overeating that they never stop to think how odd it is that the human body can make such a colossal blunder. After all, the body is usually a miracle of self-regulation. Spend all day at the gym, and you still won't end up with more muscle than you can carry around (unless you dose up on steroids). Shower every hour, and you still won't end up with skin that's six inches thick. But scarf a few dozen extra Twinkies, and you just might end up saddled with a sizable spare tire of fat.

Clearly a better-designed body would compensate for this mistake by jettisoning extra calories. After all, few humans can use an extra 100-pound energy reserve. A more intelligent body would reach a set fat threshold and then stop accumulating any more. But the human body never developed this skill because it never found itself in this situation—until now.

Tens of thousands of years ago, it was critical that our ancestors ate every calorie that came their way. Obesity—the result of being able to find and eat vastly more of the nutrients that you need to survive in a harsh, physically demanding world—was virtually impossible. (In fact, it's only in the last 50 years that scientists have collected worldwide obesity statistics—until then, obesity seemed like the relatively rare problem of a few compulsive eaters.) Your genes have been passed down from the most successful humans in those prehistoric days, and they had bodies that craved all the food they could get.

In other words, the sweeping timescale of evolution has honed your appetite-control system into a powerful force for preventing starvation—not for managing weight. If it's any comfort, blame the cavemen. If they had found a way to eat themselves to diabetes and coronary heart disease 50,000 years ago, the forces of evolution would probably have solved the problem by now.

Good Reasons Not to Diet

When trying to control our weight, many of us try to adjust our eating first, often using a too-good-to-be-true fad diet. But as you've learned, eating is just part of your body's complex weight-management system—and many of the other factors are out of your control.

If you're a potential dieter, here are the facts you don't want to know:

- Most dieters succeed initially.

- Of those who do, very few keep their slimmed-down body weight for more than a couple of years. Most regain their original weight, along with a few extra pounds of anti-starvation insurance.

- Most dieters cannot lose large amounts of weight. As they reduce their food intake, the body slows its metabolism to compensate.

- Time is the great diet equalizer. Although some diets appear to work faster at first, the results are strikingly similar over the long term.

All of these points add up to a compelling reason not to diet—namely, diets rarely work. Unfortunately, a quirk of human biology makes diets appear to work better than they actually do, but only for the first few weeks. When you first reduce your eating, your body uses its storage of glycogen first. *Glycogen* is an energy reserve made up of sugar—it's not as efficient as fat storage, so your body doesn't keep a lot of it around. However, your body can get the energy from its glycogen reserve more easily, so that's what it does. The trick is that glycogen can be stored only with a good bit of extra water. As you burn up the glycogen, your body releases the extra water. In other words, the easy initial weight loss of any new diet is usually excess water that your body will regain later. Fat is more persistent.

If all this isn't bad enough, sudden weight loss may cause problems of its own, or at least fail to give you the health boost you expect. Here are the potential problems:

- When you lose large amounts of weight quickly, you're likely to lose valuable muscle mass at the same time, especially if you aren't performing a regimen of strength-training exercises.

- Some studies suggest that extreme dieting cuts out subcutaneous fat first, leaving the more dangerous visceral fat in place.

- Going through continuous cycles of weight loss and weight gain strains your body, and may encourage it to keep hoarding calories, making your weight creep up the scale.

Lastly, a body that's been rapidly dieted down to a healthy BMI might not be healthy at all. Some controversial studies have discovered a phenomenon known as the *obesity paradox*. As you might expect, high BMIs are associated with higher rates of certain debilitating diseases (for example, congestive heart failure, coronary artery disease, and chronic renal disease). But the obesity paradox finds that once people *have* these diseases, those with higher BMIs have better survival rates. Some scientists believe this effect is just an illusion that might be caused by the fact that, at their more severe, many of these diseases can cause weight loss. But many experts argue that it reflects the danger of rapid weight-loss regimens, which can eat up healthy tissue from muscles and organs along with fat.

The Practical Side of Body Science

Sane Dieting

In the face of all this bad news, what's a calorie-compromised person to do? The best advice is to learn from the dangers of kamikaze dieting. Unless you're morbidly obese (with a BMI over 40), don't strive for the massive weight changes that get all the applause on *Extreme Makeover*. Instead, take the more cautious, careful approach outlined here:

- **Have patience.** Dramatic weight-loss is unsustainable. What's the point in a two-year plan for extreme weight loss if all you get is six months of skinniness? Instead, make lifestyle changes you can sustain forever. Focus first on preventing weight gain, and then aim to lose small amounts at a time. And don't abandon your plan after one binging disaster. Instead, expect occasional setbacks and work through them.

- **Instead of removing bad foods, add good ones.** After all, if you take all the food that nutritionists have ever criticized out of your diet, you'll be left with meals of Melba toast and oat bran (until the Atkins followers snatch that away from you, too). So concentrate on getting the benefit of good, nutritious food (as outlined on page 194). Add one new healthy food a week, and before you know it, you'll have crowded most of the bad stuff out of your stomach.

- **Don't invite extra calories into your home.** To avoid binging on junk food, keep it out of your shopping cart in the first place. Once the food is in your house, the dynamic changes from "Why should I eat this?" to "I'm obviously going to eat this—why not now?"

- **Keep eating the fat.** Fat allows your body to build essential compounds and slows the digestion of your meal, helping you feel full. Cut out all the fat, and you're likely to fill the void with something else—like a second serving of carbohydrates. Furthermore, the Nurses' Health Study—a well-respected, long-running study that tracks the health of over 100,000 nurses—suggests that there is no link between total fat consumed and heart disease, as long as you avoid the manufactured *trans-fats* found in many processed foods. This suggests that the fat on our bodies can't blame the fat in our diet.

You'll get even more insight into good food in Chapter 8, when you tackle digestion.

Winning the Battle of the Bulge

Let's be honest. If you're wrestling with weight, you've just read a lot of bad news, very bad news, and abjectly bad news. Basically, the problem is that your body doesn't want you to lose large amounts of weight. It stubbornly clings to this point of view, even though science tells us slimming down would help it work better. Unfortunately, in battles like these, the body usually wins.

But don't give up just yet. If your weight is high or creeping up, you can't afford to walk away from this battle. Instead, you need to use every trick in the book to stay healthy. Here are some useful techniques:

- **Copy successful losers.** Surveys tell us that successful dieters—that rare breed of individuals who lose weight and keep it off—have a few characteristics in common. They look at their weight loss as a long-term lifestyle change, not as a quick fix. They exercise regularly, and their exercise combines calorie-burning aerobic exercise (like running, swimming, or dancing), with muscle-building weight training. But don't be alarmed—there's no need to haul out back-straining barbells or climb into expensive gym equipment. A few light weights can do wonders for your muscle tone and bone strength, boost your metabolism, and help fend off the ravages of age. Page 72 has some tips for a good exercise regimen.

- **Pretend to be skinny.** Naturally thin people behave differently at the dinner table. They're more likely to eat before they're head-throbbingly hungry, and so they're less likely to overeat. They eat slower and with less embarrassment. They're picky, and they don't clean their plates if the meal includes food they don't really like or want. They also get up, move around, and fidget away dozens of extra calories a day.

- **Become aware.** Overeating is easy, if you don't think about it. Unfortunately, a great number of people spend a great deal of time devising ways to encourage your automatic eating habits. (To learn about them, check out the eye-opening book, *Mindless Eating*). To put the thinking parts of your brain back in control, avoid eating while engaged in other activities (watching reality television, working on your taxes, and so on). Some dieters find that a food journal—a daily log of all the food they eat—forces them to face up to what's on their plate. And one innovative study found that dieters who used a camera to take pictures of their every meal and snack were more likely to stick to their diets.

- **Get a good night's sleep.** Skimping on sleep causes your body to release the stress hormone *cortisol*, which promotes fat accumulation. To be in fine form, insist on eight hours a night.

If you struggle with weight, none of these tactics will completely conquer excess pounds. But they will help you wrest a bit more control away from your body's calorie-hoarding autopilot, and give you better odds in the never-ending battle against obesity.

3 Muscles

Muscles are your body's movers and shakers. They let you do everything from climbing a flight of stairs to chewing on a piece of beef jerky. But despite their obvious importance, most people have a rather strange way of treating their muscles. Some of us head to the gym and strain them for hours (mostly the stair-climbing muscles, not the beef jerky–munching ones). The rest of us spend more time worrying about other body parts—like the heart and the brain—and assume that our muscles will quietly adapt themselves to the demands of our day-to-day lives.

This is unfortunate, because the health of your muscles is vitally important to the health of your body. Even if your idea of a marathon is watching 12 back-to-back episodes of *Gossip Girl*, and the only powerlifting you do is moving groceries from the backseat of your car to the kitchen, you can't afford to ignore your muscles. Strong, well-maintained muscles improve your blood pressure, your bone mass, and your ability to burn calories. They also reduce your likelihood of injury and the risk factors for countless diseases.

In this chapter, you'll learn the basics of taking care of your muscles. You'll see how they attach to bones, how they grow, and why the only way to strengthen them is to tear them up ever so slightly with an often-tedious activity called exercise. You'll also get the truth about stretching, pick up the proper way to lift weights, and learn why a basic strength-training workout won't turn you into a world-class weight lifter with biceps the size of bread loaves.

Meet Your Muscles

From a biologist's point of view, a muscle is a tissue that *contracts*. Your arm muscles contract to swing a racket and propel a ball across a tennis court. Less glamorously, the muscles in your lower intestine contract to propel Sunday breakfast down your digestive tract. And if you're a woman about to give birth, your uterus muscles contract to do something truly miraculous—squeeze a brand-new person out into the world. All these examples have a common theme: when muscles contract, things *happen*.

Even if you think you're neglecting your muscles, they never truly get a moment's rest. Right now, for example, they're hard at work on a thousand small but essential jobs—pumping blood through your arteries, directing your eyes over the words in this sentence, and keeping your body from slumping over onto the floor.

Part of the reason muscles work so well is because your brain keeps them busy all the time with tiny, partial contractions called *muscle tone*. This activity ensures that your muscles are healthy and always ready to act. If your brain stops doing its job—which can occur, for instance, in the event of a spinal injury—your muscles become floppy and begin to waste away.

> **Note** The actual biochemical process that triggers a muscle contraction is quite complicated and, like all chemical reactions, it's not entirely efficient. In fact, as much as 60 percent of the energy that fuels your muscle movements escapes as heat. That's why the quickest way to warm up on a cold day is to jump up and down. It's also why your body forces your muscles to start contracting when your body temperature drops—a phenomenon we call *shivering*.

The Three Flavors of Muscle

All muscles are not created equal. In fact, your body has three very different types of muscle, each of which does a different kind of work:

- **Skeletal.** These are the muscles most people think of first. They're anchored to your bones and under your conscious control. You use these muscles to move your body parts (and to impress potential dates at the gym).

- **Smooth.** Most of these muscles are deeper inside your body, in thin sheets that line internal passages like your digestive tract, your blood vessels, your bladder, and (if you've got one) your uterus. You also find smooth muscles closer to the surface, where they can make the several million fine hairs on your body stand on end.

- **Cardiac.** You find this muscle type in just one place—your heart. Like smooth muscle, cardiac muscle is beyond conscious control, meaning you can't will it to contract or stop contracting. However, the structure of cardiac muscle is more similar to skeletal muscle than to smooth muscle, and its uncanny sense of rhythm is something no other muscle can match.

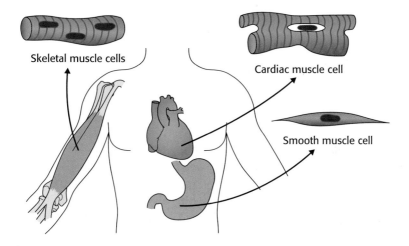

Skeletal muscle cells

Cardiac muscle cell

Smooth muscle cell

You'll take a closer look at your heart in Chapter 7, and you'll take a scenic trip down your digestive passages in Chapter 8. But in this chapter, you'll spend all your time studying skeletal muscles.

White Meat and Dark Meat

Even if your body is more Bill Gates than Mr. Universe, it still boasts about 650 skeletal muscles. Each muscle is built out of long, tough fibers that are grouped into rope-like bundles. When you throw a javelin (or just lift your posterior off the couch), these fibers pull in unison, generating the force you need.

There are actually two types of fibers at work in any skeletal muscle:

- **Fast-twitch.** These fibers contract with brief bursts of explosive force. However, they can't keep contracting for long. Fast twitch fibers are a light whitish color.

- **Slow-twitch.** These fibers contract more slowly and with less force. However, they can sustain their contraction for longer periods of time. Slow-twitch muscle fibers are reddish and darker in color than fast-twitchers because they have more *mitochondria*—the cellular factories that generate energy.

Your muscles have a different mix of fast- and slow-twitch fibers, depending on their purpose. For example, the neck muscles you use to hold your head upright perform slow, steady work all day, so you're likely to find that they have more slow-twitch fibers.

> **Note** To let this all sink in, try reflecting over a takeout chicken. The breast meat is mostly white because it's made up of fast-twitch fibers that (in better times) helped the chicken flap its way into the air with short bursts of contractions. The leg meat is a darker color because it's rich in slow-twitch fibers, which the chicken used to use to amble around all day.

If you examine the muscles of elite athletes, you usually find that their mix of muscle fibers matches the requirements of their sport. In other words, the muscles of sprinters are fast-twitch white meat, while the muscles of marathon runners are slow-twitch dark meat. Much of the difference between the amount of fast-twitch and slow-twitch fiber you have is genetic. Even hard-core exercise seems unable to change your ratio of fast-twitch to slow-twitch muscle fibers. So if you're a white-meat sort of person, all the cross-training in the world won't give you dark-meat muscles.

But don't shelve your gym shoes just yet. Most people have a wealth of underused muscle fibers. (As you'll learn on page 73, fast-twitch fibers are particularly lazy. They get involved only in extreme muscular efforts—say, lifting heavy weights.) When you exercise, you trigger a raft of beneficial changes that alter the way your muscles work in ways both straightforward and subtle. And if you train for a particular sport, your body adapts to become more and more efficient at it.

So the bottom line is this—just about anyone can develop the muscles of a proficient marathon runner, but you'll need an extra genetic gift to become a world champion.

Pairing Up

Like happy couples and dirty cops, skeletal muscles always work in *pairs*. Remember, a muscle can do only one thing—contract. It can't stretch in the opposite direction. To control your body, your muscles need to team up in sets. When one contracts, the other must relax.

Think about what happens when you bend and unbend your arm. In one direction, your biceps do the pulling. In the other, your triceps take a tug. There's no conscious way to make both muscles contract at once—if there were, the result would be a disastrous strain on your joints and possible muscle injury.

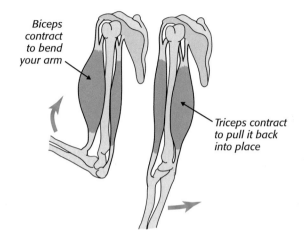

Biceps contract to bend your arm

Triceps contract to pull it back into place

Your arm bends at the relatively simple elbow joint, which acts like a hinge. More sophisticated joints allow more directions of movement. Your shoulder, for example, is a ball-and-socket joint. While your arm needs just one pair of muscles to bend at the elbow, it needs no fewer than *three* sets of muscles to rotate around your shoulder.

Fortunately, this rather complex system works automatically. (In other words, you can swing your arm around with wild abandon, without pausing to think about which muscles are doing the actual work.) However, you do need to think when you *exercise* your muscles. In fact, this is the reason that many people organize strength-training workouts around muscle groups. People who use this approach want to develop the muscle sets that work together, keeping them balanced in both appearance and function. (You'll get an introduction to strength training on page 77.)

Binding Muscles to Bones

If you were able to dissect yourself, you'd find that most skeletal muscles attach directly to bone with a tough piece of connective tissue called a *tendon*. When a muscle contracts, it pulls on the tendon, moving the appropriate bone.

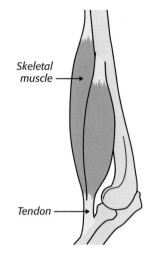

Most skeletal muscles stretch over a joint, from one bone to another. A tendon anchors the muscle on both sides.

Skeletal muscles aren't always where you expect them to be—namely, on the body part they control. The muscles that bend your *knee,* for example, sit inside your **thigh**. Similarly, the muscles that bend your elbow are inside your upper arm.

One of the more interesting examples of this quirk is found with the muscles that move your fingers, which aren't anywhere near your 10 digits. If they were, your fingers would be plump, clumsy little sausages. You'd need a jumbo-sized keyboard to type your name and a robot to tie your shoes, and you'd have no chance of playing any instrument requiring more manual dexterity than a plastic whistle.

Instead, the major muscles that move your fingers are in your *forearm*. To take a look, roll up your sleeve and watch the top of your forearm as you twiddle your fingers around. You'll see the contractions of the finger-controlling muscles in your arm. Long tendons stretch from these muscles, down your arm and through your wrist, connecting them to your fingers.

When these muscles move, they pull the tendons and jerk your fingers around, somewhat like a puppeteer controlling a marionette. Although your hand does have a few smaller muscles that help precisely position your fingers, the muscles in your forearm hold the real power.

Right now, this fact might seem like little more than a second-rate party trick. But there's a good reason to understand your tendons—they're the source of many types of injury.

For example, tendons that repeatedly rub against a narrow passage in your body can cause painful inflammation, called *tendinitis*. Tendinitis commonly involves the tendons around your knees, elbows, shoulders, or wrists.

An even more insidious wrist problem is *carpal tunnel syndrome*. It occurs when the tendons that control your fingers become chronically swollen and put pressure on your hand's medial nerve. This happens because both the tendons and this nerve share the narrow passage in your wrist called the carpal tunnel.

Even more catastrophically, you can tear or cut a tendon, which will prevent you from operating the attached body part unless you see a surgeon—fast.

> **Note** The 50-odd muscles in your face are unique—instead of using tendons, they connect directly to your skin. This attachment gives you remarkable control. It's the reason your face can communicate the subtlest of emotions, while the skin on your back lacks the same expressive power.

The Practical Side of Body Science

Caring for Your Tendons

You can injure tendons with repetitive motion and excessive exercise. Any repetitive action, from swinging a golf club to practicing a piano sonata, has the potential to cause problems, including inflammation, soreness, and pain with movement, which can range from mildly bothersome to chronically agonizing.

Here are some tips to avoid the worst problems and treat your tendons with the respect they deserve:

- **Go slow.** People often injure their tendons when they suddenly put new demands on them—for example, trying to master a new sport in a weekend or write a five-year business plan overnight. But if you build up to a new activity slowly, your tendons will grow stronger and your body can adjust gracefully.

- **Take breaks.** Almost all types of repetitive stress injury occur when you work a muscle the same way for hours at a time. If you're an office worker, take frequent pauses to stand and stretch and an hourly break to walk around the block. If you're an athlete or fitness buff, vary your workouts, try a different sport, or go for a swim.

- **If you're injured, stop.** This one is important, because if you carry on in the face of tendon pain you'll not only prolong the suffering, you'll increase the risk of permanent nerve damage. This is a particularly serious danger for hard-driving musicians (violinists, piano players, and so on).

- **If you're in doubt, see a doctor.** If the pain persists or you have other symptoms (for example, an inability to move a part of your body properly or tingling and numbness), have it checked at your doctor's office. Your doctor can rule out other problems and prescribe an anti-inflammatory medication that may help.

Exercise

Muscles are the ultimate self-tuning organs. If you rely on them to perform daily labor, they respond by growing bigger and stronger. But if you don't give them anything useful to do, they shrink. That way, your body saves on the metabolic cost of keeping them alive. After all, the more muscles you have, the more calories you burn, even at rest—and your frugal body isn't willing to waste all that valuable fuel.

In the modern world, where the hardest labor many people do is to reach across the sofa for the TV remote, our daily activity just isn't enough to keep our muscles healthy. To fill the gap, we created *exercise*—basically, a practice in which we lift heavy things and then put them down in exactly the same place, or run like crazy on a machine without actually going anywhere.

Exercise is a thoroughly modern invention. Thousands of years ago, the balance between rest and activity was almost exactly the reverse of what it is today. People spent most of their lives straining their bodies, and if they had a moment of free time, the healthiest thing they could do was rest their weary muscles. Today, we spend most of our lives sitting in one place and thinking hard (or at least trying to look like we're thinking hard). When we have time off, we use exercise to build and maintain healthy muscles—or, from a more cynical perspective, to give the illusion that our bodies are being put to good use. But if you want to spend your twilight years strong, hearty, and with all the muscular strength you need to pull your bottom off a toilet seat, you need a regular regimen of exercise.

As you probably know, there are two basic types of exercise:

- **Aerobic exercise.** This is the heart-pounding, fast-breathing sort of exercise you perform when jogging, swimming, cycling, or jumping on a trampoline. A regular regimen of aerobic exercise strengthens your heart, improves your lung capacity, and increases your endurance. It has a cascade of other potentially beneficial effects on the body, too—for example, it can improve your coordination, boost your metabolism, and burn fat. However, aerobic exercise isn't the best way to build your muscles.

- **Strength training.** This is the intense muscle-straining exercise you perform when lifting weights or doing sit-ups. It makes your muscles contract much more forcefully, but for much shorter periods of time. Although a regular regimen of strength training won't improve your stamina, it will pump up your muscles.

Understanding the Role of Oxygen

The term *aerobic exercise* actually means "with oxygen." When you perform aerobic exercise, your muscles generate the energy you need using a complex series of chemical reactions that involve oxygen. That's also the reason your breathing speeds up (to get more oxygen) and your heart begins to race (to pump that oxygen-rich blood to the muscles that need it).

Strength training is also called *anaerobic exercise*, which means "without oxygen." That's because the processes that turn oxygen into energy don't work fast enough to keep up with anaerobic exercise. Instead, your body makes up the difference with a different set of reactions that use the energy stored in your muscles as glycogen. These energy reserves are limited, which is why your muscles can't sustain anaerobic exercise for long before they seize up.

Incidentally, when you do aerobic exercise, you're primarily using your slow-twitch muscle fibers (page 67). When you do strength training, you're putting your fast-twitch fibers to work.

Both types of exercise are important, and their benefits are complementary. You'll learn more about aerobic exercise on page 172, when you explore the heart. In the following sections, you'll learn more about how muscles develop, and you'll get some practical tips to help with your own strength training.

Building Muscles

Muscle cells share one thing in common with the fat cells you met in Chapter 2—they rarely multiply. The number of muscle cells you had at birth is the same number you have now. And there's probably nothing you can do to change that.

Note There is one case when your body creates new muscle cells—if an existing muscle cell dies because of damage or disease. But this process has strict limits. For example, your body can't repair certain types of muscle tissue or patch up extensive damage, and it may fill gaps with useless scar tissue. (This is what happens if you suffer a heart attack, in which case your heart is never the same again.)

Fortunately, you don't need to increase the number of your muscle cells to gain strength—you simply need to beef up the ones you have. Oddly enough, the trick to building stronger muscles is to damage them, and the best tool for inflicting the gentle trauma you need is exercise.

Here's how it works. When you exercise, the vigorous muscle contractions create microscopic tears in your muscle fibers. As your body repairs these tears, it stuffs in a bit more protein to make the muscle a little more resilient the next time. Repeat this process over the course of a year, and you have a recipe that gradually bulks up your muscle, making it stronger along the way.

Frequently Asked Question

Should Women Lift Weights?

Many women avoid strength training because they're afraid it will give them massive, Schwarzenegger-sized muscles. But while strength training *might* cause the number on the scale to inch up (muscle is heavier than fat, after all), it certainly won't cause bulky muscles to burst out of your blouse. That's because women lack the testosterone needed to fuel serious muscle growth. If you want to be a female body builder, you can give it a shot with an extremely rigorous, extremely specialized workout, but ordinary exercisers can't get anywhere close.

So with that worry out of the way, here are some of the reasons why every woman should add strength training to her workout regimen:

- To preserve bone mass and stave off osteoporosis (page 93)
- To reduce blood pressure
- To lower the risk factors for various diseases
- To create strong tendons and ligaments, thereby reducing the risk of injury from accidents or other activities
- To boost resting metabolism (the amount of calories your body burns when it's not doing much of anything) and fight the creeping weight gain of advancing age
- To keep misbehaving men in line

Although the average woman won't lift the same load as the average man, the strength-training advice in this chapter applies equally to both sexes. That means that to get the benefits of strength training, you need to do it often enough (two or three times a week) and with the right weights (ones that are heavy enough to strain your muscles, but not so heavy that you can't control your movement). And if you aren't comfortable in the sweaty, testosterone-soaked atmosphere of the local gym, shop around until you find a more relaxed place that suits your style, or opt for a women-only fitness club.

Muscles don't necessarily need to get bigger to get stronger. Studies find that exercised muscles develop a better blood supply, which gives them improved access to oxygen and lets them work longer before tiring out. They also respond more readily to the signals your brain sends them, springing into action more easily. Exercised muscle cells also get more *mitochondria*, which are the power plants of your body. They carry out the energy-producing chemical reactions that muscles need to contract. The net effect is that an exercised body gets a larger and more easily accessible energy supply.

The Stretching Controversy

If you remember high-school gym class, you probably remember three things about it: a) changing clothes in front of adolescent peers is no fun; b) proper stretching is essential before any physical activity; and c) the meek may inherit the earth, but not before a few dodgeballs ricochet off their foreheads.

It now seems that the weight of modern science contradicts the second rule—in other words, stretching before exercise doesn't do much for your joints or muscles. In fact, *static stretching* (where you hold a stretched position for 20 or 30 seconds) temporarily weakens the muscle, and may even make injuries more likely when you stretch at the beginning of a workout. That's because resting muscle is stiff. It's just not ready to meet the demands of vigorous exercise or deep stretches. In fact, static stretching may simply increase your *stretch tolerance*, which is a fancy way of saying that your brain will allow you to extend stretched muscles a little bit more, even though they aren't yet warmed up and really ready for it.

That doesn't mean that you should leap straight from the couch onto the treadmill. Before any sort of exercise, you should perform a short 5- or 10-minute *warm-up.* This stimulates blood flow to your muscles and literally warms up your body. It also gently loosens muscles and tendons, increasing the range of motion in your joints. Your warm-up should match the exercise you're doing. If you're getting ready to run down the street, start by jogging in place. If you're lifting weights, start with a little light jogging (just enough to break a sweat) and then perform a set of your chosen exercises with a much lighter weight. Once you finish, you can begin the real muscle-straining workout.

And after your workout? Recent studies suggest that just as pre-exercise stretching doesn't prevent injury, post-workout stretching does little for muscle soreness. At one time, experts thought that stretching reduced the buildup of *lactic acid* in muscles, reducing the stiffness and soreness you feel the next day. However, the theory didn't fit the facts. Although lactic acid causes the burning feeling you experience while you strain your muscles, it quickly disappears after exercise. Muscle soreness usually appears much later (after a day or two), when there's little lactic acid left in the muscles. Today, biologists believe that the soreness is caused by inflammation, which sometimes follows the microscopic muscle-tearing that stimulates muscle growth.

> **Tip** The best ways to avoid muscle soreness are to keep up a regular exercise routine, avoid abruptly increasing the intensity of your workout, and give tired muscles sufficient time to recover (a day or two, depending on how much you exercise).

More important than stretching after a workout is carrying out a proper *cool down*—gradually stepping down your exertion until your heart rate is just 10 to 15 beats above normal. Stopping too abruptly can cause a sudden drop in blood pressure and muscle cramping.

Finally, none of this is to say that stretching isn't worthwhile—many experts believe stretching is very important for preserving flexibility and maintaining health. Good stretching regimens include yoga and tai chi, and may also involve meditation and breathing techniques that reduce stress.

Setting a Schedule

Before you decide *what* exercise to do, you need to determine *how often* you should be doing it. Exercise too frequently, and your muscles won't get the chance to patch up their tiny tears. Go too easy on yourself, and your muscles will continue to be their lazy, underdeveloped selves.

Here are some broadly accepted guidelines for strength training:

- Do a strength–training workout two or three times a week. Of course, spread your sessions out—your muscles need one to two days of rest after every session.

- Aim to do 30 minutes of strength training at a time.

- A simple strength-training regimen incorporates 6 to 12 different exercises. It's common to perform two or three *sets* of each exercise, which means you start over and repeat the whole group of exercises once you complete the first set. However, recent research suggests that a single set is enough, provided you do it perfectly (as explained in the next section).

- Use *compound exercises* that work several muscles at once. This lets you exercise more muscles more quickly, and reduces the chance of overexercising certain muscles while leaving others behind.

Note Some serious muscle-builders prefer exercises that work on specific muscles. This lets them exercise different muscle groups on different days, so they can train the whole week and still (theoretically) give each muscle enough time to heal before they exercise it again. This approach isn't appropriate for casual weight lifters, who don't have the time or skill to target just the right muscles.

- Don't abandon your aerobic exercise. To keep your heart healthy, you should perform aerobic exercise two to three times a week.

How to Lift Weights

For many people, the very *idea* of lifting weights is intimidating. If you're afraid to step up to a dumbbell because you think someone will expect you to oil up your skin and grunt like bull moose, you may be one of those people.

But don't give up just yet. The health benefits of strength training are simply too important to miss. And with a bit of up-front guidance, you'll find that weight lifting is surprisingly straightforward. In time, it can become just as comfortable as any other workout.

First, you need a solid set of running shoes, a basic set of weights (or membership in a fitness club that has them), and a bit of self-confidence. Next, you need to understand the sort of exercises you'll do—mostly compound exercises that work muscles throughout your body. Finally, you should study the following sections, which explain the basic principles of successful strength training. Best of all, there's no grunting required.

Aim for the point of exhaustion

Proper strength training works only if you challenge your muscles—in other words, when you lift a weight that's not easy to lift. Beginners sometimes make the mistake of opting for lighter weights and longer workouts. This sort of workout may boost muscle endurance, but it won't cause the microscopic tears that spur muscle growth. Of course, choosing a weight that's too heavy is even worse, because it can cause injury. So how do you choose the right weight? The answer lies in understanding the *8/14 rule*.

Every strength-training exercise involves repeating the same action several times in a row. This is called a *set*. Ideally, you repeat the exercise 8 to 14 times. The right weight is one that's light enough for you to lift comfortably and stay in control, but heavy enough that you're completely tapped out by the last repetition. This lets you work your muscles to the *point of exhaustion*.

Tip It takes a bit of experimentation to determine the weights you should use for each exercise. If you feel like you can keep going after 16 repetitions, you probably need a heavier weight (assuming you're performing the exercise with perfect form). If you can't manage eight without pausing or rushing, you need to go lighter.

Use good form

The secret to getting a good workout with weights is being fanatical about good form. It doesn't matter whether you're a championship weight lifter with the body mass of a middle-aged rhinoceros or a computer genius who hasn't left his home office in years, the rule is the same. Lift weights sloppily and you're more likely to cause an injury. But concentrate on carefully directed motion, and your muscles will get the maximum benefit in the minimum amount of time.

The key is *control*. If you find that you're beginning to lose control over the weights you're lifting—dropping them instead of lowering them slowly—you're lifting too much. You should be able to raise and lower your weights at the same speed for each repetition, including the last one.

This rule isn't just a safeguard against injury. It's also a practical guideline that helps you get the most from your workouts. When you lower a weight, your muscle performs an *eccentric contraction*, which means it produces force as it *lengthens*. (When you lift a weight, your muscle performs a *concentric contraction*, exerting force as it shortens, or contracts.) Modern-day exercise science suggests that eccentric contractions cause more of the microscopic muscle tearing that stimulates muscle growth.

Here are a few more tips to keep in mind when you approach an exercise:

- **Relax.** Concentrate on slow, efficient movement. Never swing a weight or use momentum to lift it. Lowering a weight should take more time than lifting it.

- **Breathe.** Exhale when you lift or push the weight, and inhale as you're bringing it back to your starting position. Don't hold your breath during an exercise.

- **Stand up straight.** Pay attention to your posture, keep your balance, and hold in your abdominal muscles.

- **Pain is not OK.** Not even if you like it. If you find that raising or lowering a weight past a certain point causes pain, don't push it that far (or choose a different exercise). You should perform each exercise through the greatest range of motion you can achieve without pain.

- **Listen to your body.** At first, you'll lift less weight than you might expect. Conversely, as you become stronger, you'll need to increase the weight to maintain the same level of difficulty. Otherwise, your muscles will have a picnic.

- **If you're not sure, get help.** The best option is to have an experienced professional show you how to do an exercise—for example, a personal trainer at a fitness club. If you can't do that, search for an instructional video from a reputable fitness website that shows how to do an exercise. (You'll get a few such links when you try out some strength exercises later in this chapter.)

Always warm up

Don't go in cold. Warming up is essential before any type of exercise to prepare your muscles and prevent you from hurting yourself. A good weight-lifting warm-up starts with 5 to 10 minutes of light aerobic exercise (for example, running in place, jumping up and down, and trying not to look ridiculous). It's also a good idea to do a warm-up set before you start each individual exercise. For your warm-up, perform the same exercise but with a much lighter weight (or no weight at all).

How Muscles Age

For the average person, age 25 is the high watermark for muscle mass. After that, muscles begin to waste away, shrinking at least 10 percent before age 50, after which the reduction quietly picks up speed. In the years after 50, adults lose an average of one-half to one pound of muscle each year—but often don't notice the loss because fat slides in smoothly to replace it. The biological term for this phenomenon is *sarcopenia*. If ignored, it can leave older people unable to carry out daily activities and make them defenseless against injury. An otherwise minor accident—like a fall in the bath—becomes more likely, more damaging, and more difficult to recover from.

Many of the factors that cause sarcopenia are biological. As the body ages, it becomes more efficient in reclaiming unused muscle, and slower to recover after a bout of exercise. The body's levels of testosterone and human growth hormone plummet. Muscle growth slows, and some muscle fibers may die off completely. However, an equally important factor is the changing lifestyle of midlife and old age. Without continuous activity, muscles shrink, much as they do for astronauts after mere weeks of a weightless space flight.

Fortunately, there's no need to go gently into that good night. Studies consistently show that older adults can use strength training to maintain their muscles and stimulate new growth in more or less the same way that young people can. In fact, two strength-training sessions a week is enough to keep your muscles in tune and your body in good health for years to come. (However, as you age, your body becomes less forgiving of injuries and strain. So if you embark on a new workout after 60, it's a good idea to get your doctor's clearance, join a workout group, and get the advice of a personal trainer.)

A Grab Bag of Strength Exercises

Now that you know how strength training works on your muscles, you're ready to design your own personal workout. And while the world has more exercises than flavors of ice cream, you can get a surprisingly good newbie workout with just a few classics. The trick is to pick good-quality exercises that work many groups of muscles at the same time, and to perform them with flawless form.

The following sections introduce six essential strength exercises. You can use these as the foundation for a simple strength workout, or you can use them as a starting point before you progress to more ambitious, varied workouts. (On page 87, you'll get some tips for stepping up to the next level, including links to websites with sample exercises.)

Before you get started, here are a few quick review points:

- Perform each exercise 8 to 14 times, with perfect form.

- Start with light weights (or no weight at all) until you master the move, or longer if you find you're getting a decent workout. If you want to add just a little weight, you can hold a light object in each hand (say, a book, a soup can, or a bottle of water).

- If in doubt, check with your doctor before you begin any sort of new exercise. This is particularly important if you have an injury, an illness, or a pre-existing medical condition (such as high blood pressure).

1. Squats

It may have an unglamorous name, but this basic exercise is the favorite of personal trainers everywhere. In fact, squatting is often described as the most effective exercise move of all time. Its appeal lies in the fact that it exercises just about every muscle in the lower half of your body.

If you've ever watched championship weight lifting (say, in the Olympics), you've seen extreme squatting. Basically, it goes like this—absurdly muscular individuals attempt to stand up with barbells of enormous weight pinned on their backs. While any self-respecting personal trainer can help you do a safe, unintimidating squat, most beginners prefer to start with a modified version of the exercise that doesn't seem quite so threatening (and doesn't involve standing under a heavy object).

Here's how it works:

1. Stand with your feet slightly more than shoulder-width apart. Hold a dumbbell in each hand. (Or, do it without any weights the first time.)

2. Crouch down, while keeping your back straight and breathing in. Stop when your thighs are parallel to the floor (as though you're sitting in a chair). Don't let your knees move forward past your toes.

3. Pause, then stand up again, slowly. As you do, look forward and breathe out. This is the step where you really exert yourself.

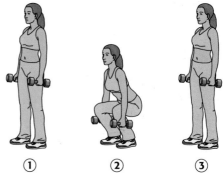

① ② ③

2. Lunges

Lunges are another wildly popular exercise for working out the muscles in your lower body. Because a lunge involves more movement than a squat, it helps you develop functional stability, balance, and overall body coordination.

Performing a lunge is sort of like taking a big step forward, then moving back to the original position. As always, proper form is the key to ensuring that you're not putting unnecessary stress on your knees. Practice this one before adding weights and, if in doubt, review a demonstration video like the one at *www.mayoclinic.com/health/lunge/MM00723*.

Here are the steps to follow:

1. Stand up straight with your arms hanging at your sides and (option- ally) a weight in each hand. This is the same as the starting position for squats.

2. Pick a leg. Take a big step forward with that leg, breathing in and bend- ing the knee until your thigh is parallel with the floor. (Be careful not to lean forward or let your knee slip in front of your toes.) While stepping forward, bend your rear leg and raise the ball of your rear foot off the ground.

3. Pause, then move back to your original position by pushing back with the front leg, breathing out.

4. Repeat the exercise by step- ping forward with the other leg.

The lunge is a perennial favorite, and it has countless variations. Lunges are also performed without weights (and for longer periods of time) as part of many high-intensity workouts.

①   ②   ③

3. Overhead Press

The overhead press is a compound exercise that targets your upper body, making it a natural complement to the squat. It concentrates on the shoulders, but also involves muscles in the back, arms, and abdomen, all of which act to stabilize the movement.

> **Note** In the variation described here, you do the exercise standing, but there are alternatives. If you perform overhead presses in a chair that has a firm, supportive back, you can lean back and lift heavier weights. This suits power lifters and demands more work from the shoulder muscles, but it also reduces the workout that the stabilizing muscles get in the rest of your body.

Here's how to perform this exercise:

1. Stand with your feet slightly more than shoulder-width apart. Hold a dumbbell in each hand and bend your arms at a right angle, so that your forearms are pointing straight up and your elbows are pointing down.

2. Breathe out as you push the weights up slowly, straightening your arms. Keep your back straight and continue looking forward. Make sure that the weights don't drift forward or backward as you raise them, and don't let them touch each other when they're above your head.

3. Breathe in as you slowly lower the dumbbells to their starting position.

To extend the arm-muscle benefits of this exercise, you can tack a bicep curl onto the beginning. First, start with your arms hanging down at your sides. Then, bend your arms pulling the weight toward your shoulders. Next, turn your palms out and raise your arms to the starting position for the overhead press. Finally, complete the overhead press exercise. This requires a bit more coordination, so don't try it until you're thoroughly comfortable with the basic overhead press.

> **Note** Need some help? You can get pointers from a quick instructional video at *http://befitandstrong.com/shoulder-exercise-overhead-dumbbell-press*.

4. Bench Press

The bench press is another upper-body exercise, but this one requires more work from the muscles in your chest. In this variation, you'll perform it with two dumbbells (rather than a single barbell). This forces your muscles to work a little harder to stabilize the weight you're lifting.

Here are the steps:

1. Lie flat on your back. Use a weight bench if you have one, or lie flat on a floor mat if you don't. If you're on the floor, bend your knees and place your feet flat, as if you're about to perform a traditional sit-up. If you're on a weight bench, bend your knees over the edge of the bench and place your feet flat on the floor.

2. Begin by holding the weights close to your chest. Breathe out as you push them up, continuing until your arms are almost completely straight. The weights will be close together, but shouldn't touch.

3. Breathe in as you slowly lower the weights out to your sides. Continue until your elbows are even with your shoulders, but no farther. If you're on a weight bench and you let your elbows drop below your body, you'll put unnecessary strain on your shoulder.

Tip To see the bench press in action, with some helpful pointers, watch the demonstration at *www.mayoclinic.com/health/chest-press/MM00738*.

One of the nice things about the bench press is that it's only one of several exercises you can do while lying on your back. For example, you can follow up with a pullover, where you lower a weight behind your head (see *http:// exercise.about.com/od/exerciseworkouts/ss/backexercises.htm* for instructions). Or, if you're completely spent, try the basic back-strengthening exercise described on page 103.

5. Abdominal Crunch

What would the world be like without this exercise? It's a modified sit-up that's every gym teacher's favorite torture tool and the first line of defense against a paunchy midriff. It's also far friendlier to back muscles, which can be strained by improper sit-ups.

Technically, the abdominal crunch isn't a weight-bearing exercise. It's often called a "core-strength" exercise because it exercises the core set of abdominal and back muscles that promote good posture and reduce your risk of injury. But because you don't choose the heftiness of the weight you're lifting when performing an abdominal crunch, you need to deviate from the standard 8/14 rule.

That doesn't mean you should stomach-crunch the day away. As you become a skilled cruncher, you should strive to perform two or three sets of 25 or fewer crunches. The goal is most decidedly *not* to perform as many crunches as you can. If you can breeze through 50 without a care in the world, you need to get fanatical about good form, find ways to intensify your effort, or switch to a more difficult variation (such as doing abdominal crunches on an exercise ball).

Here's how to perform a basic crunch:

1. Lie down flat on your back. Bend your knees, keep your feet flat on the floor and hip-width apart, and place your hands across your chest.

> **Note** Some crunchers put their hands behind their heads. Although this isn't necessarily wrong, it increases your risk of injury—if you get tired or your form gets sloppy, you can easily strain your neck.

2. As you breathe out, contract your abdominal muscles as tightly as you can, and slowly raise your head and shoulders just a few inches off the floor.

3. Hold your raised position for a few seconds, breathing in and out continually, and contract your stomach as hard as you can.

①

4. Slowly lower back down, but don't relax all the way. This is a key point to getting the most out of a crunch—if you simply drop your body back down under the effect of gravity, you'll lose half the benefit of the exercise.

②

To get the best results out of your abdominal crunch, concentrate on making sure that your abdominal muscles are doing all the work. Resist the urge to push with your feet, pull up with your arms, or lead with your chin. You can watch a basic instructional video at *www.mayoclinic.com/health/abdominal-crunch/MM00725*.

6. Push-Ups

Although push-ups seem old-fashioned, they're a surprisingly practical compound exercise that requires no extra equipment. Importantly, push-ups don't just tax the muscles in your arms and chest, they also require muscles throughout your body to stabilize your motion. And if you're one of the many new exercisers who can't complete a single push-up, there are two practical variations you can start with: wall push-ups and bent-knee push-ups (shown in the picture here).

Here's how to do a classic push-up:

1. Lie chest-down on an exercise mat. Put your hands, palms down, on either side of your body at shoulder level. Keep your feet together, toes down.

2. Straighten your arms to push yourself up, breathing out. Keep your back straight, your abdominal muscles tight, and your toes on the floor.

3. Pause, then lower yourself until your chest nearly touches the floor, breathing in. If you can, push yourself back up and continue with another repetition.

If a standard push-up is too hard (as it is for many), start with a bent-knee push-up. Before you begin, while you're lying face-down, bend your legs into a right angle and lift your feet off the ground. (You can watch a demonstration at *www.mayoclinic.com/health/modified-push-up/MM00735*.) If that's still too taxing, try a wall push-up. First, stand and face a wall. Then, lean against the wall and place your hands on either side at shoulder level. Finally, push against the wall. It's essentially the same exercise, except now gravity is on your side.

Even Better Strength Training

The six strength-training exercises in this chapter are a great start. But if you've out-grown them, or if you want more guidance, there are plenty of resources to turn to. Here are some excellent options:

- **Explore the Web.** There are plenty of books and videos that dissect popular exercises, but the Web is the unchallenged champion of free and comprehensive workout information. An excellent start is the sprawling exercise and fitness pages on About.com—dip your toe in at *http://tinyurl.com/4nmlj*. The Mayo Clinic provides a smaller set of articles and some helpful videos at *www.mayoclinic.com/health/fitness/SM99999*.

- **Find a gym.** Everyone can benefit from a well-stocked gym with like-minded fitness buffs and a supportive staff. You need to shop around to find one that suits you. Keep in mind that the obvious criteria (price and location) aren't the only factors. To make sure you'll feel comfortable, check out the crowd (Lycra-clad treadmill bunnies? Macho men grunting like bulls?). And make sure you can make use of the most important gym benefit—the help of a personal trainer.

- **Personal trainer.** You don't need to have one on staff. A monthly consultation is enough to help you test your fitness level (and see how it changes over time), devise a workout regimen that suits you, and get pointers to demystify tricky exercises. If you're ready to invest in this experience, make sure you go to a personal trainer that has a degree in health science and certification from an internationally recognized organization, such as the NSCA (National Strength and Conditioning Association).

4 Bones

I f your body were nothing more than a loose package of skin, muscle, and fat, you wouldn't be much to look at. Fortunately, your body includes a sturdy substructure—your *bones*—that shapes you into the fine-looking specimen you are today.

You probably think you understand your bones pretty well. They support your body, protect your tender organs, and provide a handsome frame on which to drape some designer duds. But like the rest of your body, your bones are doing a lot more than you may realize. For one thing, your bones are a work in progress—and they stay that way for your entire life. That's because bones are built out of *living tissue* that continuously refashions itself. Your body constantly creates small holes in your bones, then patches them up—this keeps your bones healthy and strong. And, like your muscles, you strengthen your bones by straining them in just the right way.

If you think these facts don't add up to much more than fodder for a late-night biology bash, think again: Understanding these processes gives you the essential insight you need to care for your body.

In this chapter, you'll see how this knowledge can help you sustain sturdy bones, avoid joint pain, and keep your back on track. In fact, your new-found awareness just might make the difference between a body that glides gracefully into old age and one that limps into the hospital for double hip-replacement surgery.

The Skeletal System

Skeletons bring strange images to our minds. We associate them with long-dead bodies, Halloween parties, and pirate flags, which is why it seems odd to imagine a skeleton engaged in ordinary, day-to-day activity. Consider, for instance, this decidedly relaxed skeleton, drawn as a study in anatomy by Mecco Leone in the 19th century:

We don't know what this skeleton is doing—perhaps it's cleaning its foot, scratching an itch, or searching for a wart (which it certainly won't find). Whatever the case, the picture is a vivid reminder that skeletons aren't simply symbols of death and decay. They're a fundamental component of the human body—one that lurks just beneath our more familiar layers of skin, fat, and muscle.

> **Note** Your skeleton has a set of 206 neatly interlocking bones. Infants, however, have many more bones—in fact, they start with an impressive 350 bones at birth. As they grow, their bodies replace soft cartilage with new bone matter, and some bones fuse together. For example, a child's skull begins as six separate bones that gradually fuse and harden.

The Benefits of Bones

The purpose of your skeleton seems simple enough. After all, if you didn't have one, you'd be as saggy as a beanbag chair and a lot less fun. But as you'll discover, your skeleton isn't just a handy place to hang your body. It's actually a hard-working team member that helps you out in several ways:

- **Protection.** Your bones provide rigid pieces of body armor that protect your critical squishy parts. The best examples of this defense are your skull (which wraps your brain in a non-removable helmet), your spinal column (which sheaths the central highway of your nervous system), and your breastbone and rib cage (which deflect damage away from internal organs like your heart and lungs).

- **Movement.** As you saw in Chapter 3, your muscles are your body's real movers and shakers. But muscles need to act *on* something to produce movement, and that "something" is your bones. Your muscles pull them like the levers of an intricate machine.

- **Storage.** Your bones are a reservoir of minerals like calcium, which your nervous system and muscles need to function. For example, if your blood doesn't have a bare minimum of calcium, calcium will gradually leech out of your bones and into your blood. (This happens during the ongoing process of bone repair, which you'll learn about on page 92.) Some bones also store a last-ditch deposit of fat, which they keep in their hollow core.

- **Production.** As you'll see in the next section, bones don't just lie there lifelessly. They house a busy factory that creates blood cells.

> **Note** *Blood cells* are tiny cells that circulate in your bloodstream. They play several essential roles. *Red blood cells* carry life-sustaining oxygen throughout your body (page 136). *White blood cells* produce antibodies that battle infection (page 207). *Platelets* create clots that prevent your blood from leaking out of damaged skin. Without any one of these players, you'd be in serious trouble.

Men and women have subtly different skeletons. On average, women have narrower shoulders, shorter arms, and wider hips than men, so women are less suited to actions like throwing and running, but have better stability and a lower center of gravity. There are also a number of subtle differences between male and female skulls. Women's skulls, for example, tend to be less angular, and they have a less pronounced, softer chin. In a random face-recognition test, most people can quickly separate the women from the men.

Living Bone

If you're like most people, you think about bones as the dry, stiff specimens you see holding up dinosaur heads in museums or keeping turkey legs together on Thanksgiving. But the bones in your body are a world away from these fossilized and cooked leftovers. They're still tough and impressively resilient, but they're built of healthy, living tissue, just like all the other parts of your body. That's why children can grow taller month after month, and why your body can restore a cracked bone to its original strength. On the darker side of things, that's also why bones are susceptible to cancer.

You might assume that bones bother changing themselves only at certain times in your life—for example, as you grow through childhood or recover from an injury. But even if you're an adult with pristine bones, your body is constantly busy breaking them down and rebuilding them in a process called *remodeling* (which, happily, doesn't involve unreliable contractors or flighty home decorators).

To perform this ongoing regeneration, your body dissolves small patches of bone one bit at a time, and then fills the resulting holes with new material (sort of like an industrious road-repair crew). Remodeling serves three purposes. First, it fixes microscopic cracks in your bones. Second, it rebalances the strength of your bones, toughening up the parts that you strain most often. And third, it manages your body's level of important minerals, like calcium.

For example, if you have extra calcium circulating in your blood, your body incorporates it into newly laid-down bone. But if your body is short of calcium, it reclaims the mineral from the chewed-out bone matter as it remodels, and contributes less calcium to the newly patched-up bit. This presents a problem, because without those hard minerals, your bones won't have their full and proper strength. That's why a low-calcium diet puts you at increased risk of developing fragile bones.

Remodeling is slow work—in fact, it takes months to fill a single hole. However, you have roughly a million independent repair crews working throughout your body, and together they do enough work to give you a new skeleton every decade of your life.

Bone Health

Remodeling presents another potential problem. When your body restructures your bones, it uses one set of cells to break down bone matter and another to rebuild it. If the wrecking crew starts working more quickly than the repair crew (which, after all, has the more difficult job), the whole process weakens your bones. Left unchecked, this leads to brittle bones and a dangerous condition known as *osteoporosis*.

Osteoporosis is a dismayingly common condition in the over-50 crowd. It's particularly common in women after menopause, because the change in hormones affects the body's rate of bone construction. Osteoporosis usually develops without symptoms until your bones are severely weakened, at which point you can fracture them from seemingly benign events, like a negligible fall, a minor bump, or even an ordinary strain like coughing.

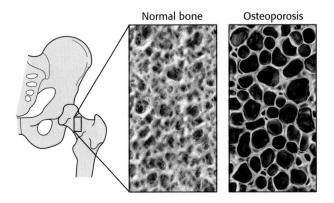

Normal bone   Osteoporosis

Osteoporosis has the potential to greatly reduce the quality of life in your later years. However, it's highly preventable if you follow the right steps throughout your life. Here are the cornerstones of good bone health:

- **Exercise.** Like the rest of your body, your bones need to be used. In fact, they grow denser and stronger in the areas you stress most. Weight-bearing exercise is best for bone health (see page 77), but even aerobic exercise (running, jumping, hip-hop dancing) helps, because these activities strain your bones with a force equal to several times your weight.

- **Calcium.** It's a critical nutrient early in life, when your bones are growing. It's also essential later in life to ensure that your body doesn't open up your bone checking account and make a calcium withdrawal. Adults from the ages of 19 to 50 should strive for at least 1,000 milligrams of calcium per day. At the same time, make sure you're getting enough vitamin D (page 13), because it aids in calcium absorption.

> **Note** The easiest way to get your calcium is from skim milk, which supplies about 300 milligrams per glass. Other calcium-rich foods include cheese, yogurt, and fortified cereal. Some vegetables, nuts, legumes, and fruits provide smaller amounts of calcium—usually closer to 100 milligrams per serving. (You can read a detailed calcium food ranking at *http://tinyurl.com/blnwya*.) Avoid mixing calcium-rich foods with caffeine, because it impedes calcium absorption. Finally, if you don't think you can consistently get the calcium you need from your diet, or you're at particular risk for osteoporosis, see your doctor for advice on taking a calcium supplement.

- **Smoking.** Avoid it. (No surprise there.)
- **Testing.** If you're concerned about the health of your bones, you can take a simple, safe test that scans your body with a low-dose x-ray. The results will tell you if you're suffering from the early stages of osteoporosis. In addition, the test will give you a benchmark against which you can compare future results to see how your bones change over time.

The Blood Factory

Learning that your bones are made of living tissue may surprise you, but your bones have even more going on. This time, however, the party's on the inside.

To understand what takes place, you need to realize that bones aren't solid all the way through. Instead, your bones have a hard outer layer and a honeycombed inner layer that's laced with tiny holes. This combination gives bones an interesting characteristic—they're both impressively strong and surprisingly light. In fact, for their weight, bones are one of the strongest materials on earth.

Furthermore, many bones have a hollow cavity packed with jelly-like *bone marrow*. Bone marrow comes in two flavors: yellow marrow, which simply stores fat, and red marrow, which creates new blood cells. Thanks to the red marrow, your bones are nonstop assembly lines, churning out more than 100 billion blood cells *every day*—important work indeed.

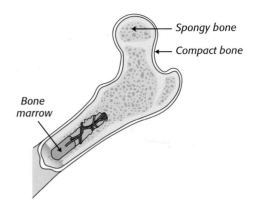

In fact, a bone somewhere in your body created every blood cell that's circulating in your body right now.

> **Note** Yellow marrow is found in the middle of long bones, like those in your legs. Red marrow is found primarily in flatter bones, like your hip bones, breast bone, skull, and shoulder blades.

Clearly your bones deserve more respect than they usually get. So the next time you see a skull and crossbones slapped on a bottle of lethal toilet-bowl cleaner, take a moment to reflect on your bones and their life-giving role as your body's bottomless blood bank.

> **Note** Bone marrow doubles as a supremely nutritious food delicacy (at least when it's from some other creature's bones). Packed with protein and unsaturated fat, it's a key ingredient in Vietnamese *pho* soup and the highlight of Italian *osso bucco* (braised veal shanks). Some anthropologists even believe that marrow was a staple in the diet of our distant ancestors. The theory is that these early humans were inept hunters who rarely caught a good meal—so they spent most of their mealtime cracking open the bones that more capable predators left behind.

Joints

On their own, bones are a remarkable piece of biological machinery. But the true miracle just might be the way a skeleton that's stronger than concrete still allows the supple movements of a ballet dance.

Joints are the means by which one bone glides, rotates, or bends around another. They ensure that your bones fit snugly together while allowing them an impressive range of movement. Different joints allow different types of movement—for example, joint limitations explain why you can't spin your hand around your wrist or fold your head back until your ears rest on your shoulder blades. The image here shows four different joints, from the relatively simple folding joints at the knees and elbows to the more versatile ball-and-socket joints in your shoulders and hips.

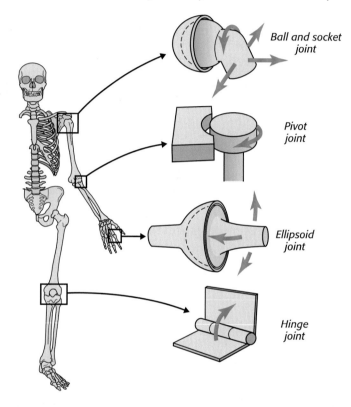

Ball and socket joint

Pivot joint

Ellipsoid joint

Hinge joint

Joints are a surprisingly complex part of your skeleton. The bones in a joint are padded with cartilage, braced with ligaments, and lubricated with a special fluid. Here's how each part works:

- **Cartilage.** This tough, flexible tissue pads the end of the two bones that meet in a joint. Cartilage helps cushion and protect the joint, absorbing shocks and preventing friction. (Cartilage also supports other structures in your body—for example, it creates the shape of your ears and the tip of your nose.)

- **Ligaments.** Ligaments are sturdy bands of elastic tissue that connect one bone to another. Your body often uses them to support a joint—for example, four major ligaments hold your knee in place and connect it to your leg bones.

- **Synovial fluid.** This thick, stringy fluid fills many of your joints. It reduces friction between pieces of cartilage and other tissue in the joint, which helps reduce wear and tear.

> **Note** Synovial fluid is probably also responsible for the exquisitely annoying sound of knuckle cracking. The thinking is that over time, bubbles of gas accumulate in the synovial fluid of your joints. When you pull your finger ever-so-slightly out of place, the knuckle's bones separate, and the tiny bubbles combine and burst in a distinctive pop. In theory, this practice could eventually cause loss of grip strength or joint pain, but so far no study has found knuckle cracking to cause anything more dangerous than a seriously annoyed spouse.

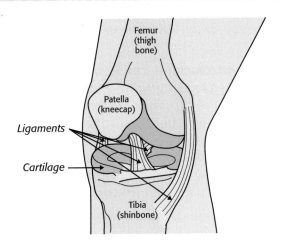

Incidentally, "double-jointed" people don't have extra joints, just abnormally expanded joint movement for some other reason. They may have misaligned joints, for example, or abnormally shaped bones, or weaker ligaments and tendons. Alternatively, they may be less sensitive to the signals that indicate when the bones of a joint are out of place. Whatever the case, their condition puts them at higher risk of bone injury and arthritis.

Arthritis

Arthritis is a group of conditions that cause joint problems, and they becomes increasingly common in the later stages of life. In the most common type of arthritis, *osteoarthritis*, the joint cartilage gradually wears away, allowing the bones to grind against each other. The causes aren't clear, but heredity, excessive weight, old age, lack of physical exercise, and injury all seem to play a part. If you have joint pain or stiffness that lasts for more than two weeks, it's best to check it out with your doctor.

Although doctors can't cure osteoarthritis, you may be able to avoid it—or at least delay its onset. The best advice is to adopt a regular routine of strength exercises (page 80). You should also avoid straining your joints with repetitive motions or dramatic changes in the way you use them. (For example, don't expect your body to make the jump from weekday office worker to long-weekend football superhero.)

Your Spine

Of all the bones in your body, few are as important as the stack of 33 that forms your spine.

Your spine (or *vertebral column*, as it's known to biologists and Trivial Pursuit players) is the centerpiece of your skeletal system—your body's equivalent of the center pole in a circus tent. It anchors your body, allowing you to stand upright, support your head and arms, and keep your balance. It's a central attachment point for countless muscles, and it encases your critically important *spinal cord*, which ferries commands from your brain to the far regions of your body (and sends sensory information in the reverse direction, from your body parts back to your brain).

Your spine is also one of the easiest parts of your body to injure—poor posture, repetitive actions, or heavy lifting can easily drive you to the medicine cabinet or leave you lying helplessly under an office table. In fact, back pain affects 80 percent of adults at some point, and nearly half of them suffer at least one bout of back pain every year. In the U.S., lower-back pain is the fifth most common reason for doctor visits.

Part of the reason your spine is so vulnerable is its complexity. Of its 33 bones, 24 are bony disks called *vertebrae*. Every two vertebrae form a separate joint, each with its own ligaments holding it together and its own tendons binding it to the appropriate muscles. This design lets you freely bend and twist your spine. Your body limits your range of motion not with the ligaments themselves, but with the bony protrusions of each vertebra, which lock together if you try to bend them past a certain point. This important safeguard prevents a strained back from causing life-threatening paralysis.

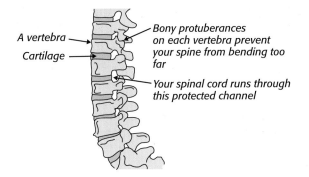

A vertebra

Cartilage

Bony protuberances on each vertebra prevent your spine from bending too far

Your spinal cord runs through this protected channel

Between each pair of vertebrae is a spongy disk of cartilage. These disks become temporarily more compressed throughout the day, and permanently more compressed as you age. They're the reason you go to bed a little bit shorter than when you woke up, and why you're likely to lose an inch of height by the time you reach old age. They're also the source of more serious, long-term back ailments. For example, the infamous "slipped disk" occurs when you inadvertently crush a piece of cartilage out of shape, making it bulge into a nearby nerve.

Posture and Pain

Your back puts up with a great many indignities. Ordinarily, your spine adopts three natural curves that help distribute your body's weight. They go by the names *cervical*, *thoracic*, and *lumbar* curves.

No matter how good your posture, when you stay in a single position for a long time (for example, when you stand in line, sit at a desk, or drive across the country), you put increased pressure on parts of your spine, and you usually end up flattening or excessively curving these parts of your back. Your muscles strain to compensate, causing problems that range from a stiff neck to an aching lower back.

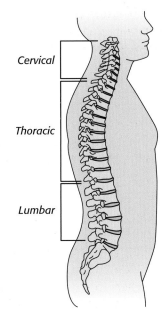

Cervical

Thoracic

Lumbar

There's no ironclad defense against back problems. However, you can do a lot by observing good posture, practicing back-strengthening exercises (see page 103), and avoiding back-straining experiences.

The first rule of good posture is to *vary your position*. Contrary to what you might think, good posture doesn't involve locking your back into an ideal arrangement for hours at a time. In fact, the human back hasn't quite caught up to the strains and stresses of modern-day activities like sitting in motor vehicles and doing office work. You'll spare your spine a lot of pain by interrupting long, tedious activities (driving, typing, and so on) to take a quick stretch and have a walk around. If you can, limit yourself to 20 minutes of focused office work or an hour of uninterrupted driving before you take a quick refresher.

The next step is to learn an invaluable skill: recognizing poor posture. If you can catch yourself when you start to slouch, you can stave off soreness, stiffness, and injury. The following sections show you how.

Standing posture

Standing is a seemingly simple activity that the human body doesn't do very well. Our bodies are designed to handle much more ambitious activities (climbing hills, hunting elk, and so on). Keep us stuck in one place, and we suffer—our backs ache, our posture slips, and we end up with more spinal distress than a performer in Cirque du Soleil.

During short periods of standing, you can reduce strain by avoiding the worst mistakes. Keep your head, shoulders, and feet lined up (imagine a straight line that links the top of your head to the middle of your feet). The biggest backaches start when you pull your head forward or lean your upper body backward.

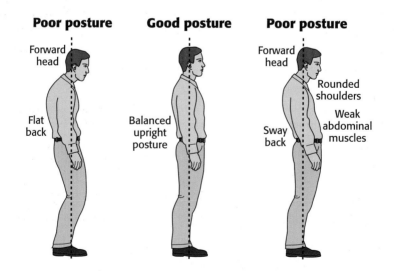

If you must stand for a long time, shift your weight from one foot to the other or rock your feet from heel to toe. Putting one foot on a very low rail, like the sort you find at some bars, can also reduce the pressure on your back. Posture experts say leaning is acceptable, but only if you put your lower back, shoulders, and the back of your head against a wall—a position that might draw a few curious stares.

Sitting posture

It's often said that the rear end is the most heavily used part of the human body. If you're an average person living in the modern world, you spend a lot of time pressing your behind into soft, yielding pieces of furniture. In fact, you probably spend more time sitting down (whether it's working, eating, driving, watching television, or reading fine books like this one) than you spend in virtually any other position.

With all our experience sitting, it's a bit disappointing to learn that most of us aren't very good at it. Proper sitting is surprisingly hard work, especially when you add computers, telephones, and office work into the mix. Part of the problem is that sitting "at the ready" quickly becomes tiring, and good posture requires constant vigilance. One funny Web video later, and you'll be slouching into your computer screen.

The first piece of advice for long-term sitters is to take frequent breaks. (You can find some simple stretches and exercises you can do without leaving your desk at *http://backandneck.about.com/od/exercise/ig/Stretch-at-Your-Desk*.) The second piece of advice is to conduct regular "posture checks." With each posture check, your goal is to review what you're doing and pull your body back into line. The most common mistakes include pushing out your stomach (which curves and overextends your back), hunching your shoulders, and craning your neck. To make sure you're sitting pretty, use the picture on this page as a quick posture checklist.

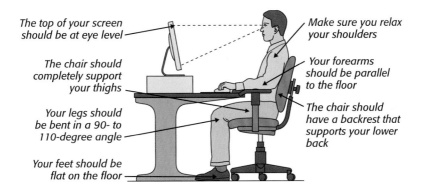

The top of your screen should be at eye level

The chair should completely support your thighs

Your legs should be bent in a 90- to 110-degree angle

Your feet should be flat on the floor

Make sure you relax your shoulders

Your forearms should be parallel to the floor

The chair should have a backrest that supports your lower back

Lifting posture

One of the easiest ways to injure your back is through careless lifting. A proper lifting technique looks a bit like the squat exercise you saw on page 81. Like the squat, it involves crouching down and using the strong muscles in your legs to do the brunt of the lifting, minimizing the stress you put on your back and reducing the amount you bend your spine.

A little preparation can make all the difference in proper lifting. Before you grab the object in front of you, adopt a wide, crouched stance. Make sure your feet rest firmly on the floor and that your back is straight, not hunched. Stand slowly, avoid twisting to the side, and concentrate on pushing with your legs, rather than pulling with your back.

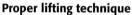

Proper lifting technique   **A very bad idea**

> **Tip** If you struggle with back pain, experts will tell you to use proper lifting for any object below waist height—even a dropped pencil.

Sleeping posture

We're not about to crawl between the sheets with you and prescribe stiff rules about how to position your sleeping body. However, if you already suffer from bouts of back pain (whether caused by long hours of office work, heavy lifting, or pregnancy), you might find that sleeping in your regular positions is about as comfortable as having a prickly pear in your pajama bottoms. In this situation, you can try two tricks.

First, if you're a back sleeper, place a pillow under your knees. This takes the strain off your lower back. (If you want to adopt this position permanently, look for a wedge-shaped specialty pillow at a pharmacy or health store.) Second, if you sleep on your side, put a pillow between your knees. Again, this helps keep your spine in the straight, neutral alignment it likes best.

Tip If you're still not getting any relief for a sore back, consider the pelvic-tilt exercise discussed in the next section. Not only can it relieve back pain, but it doesn't require you to climb out of bed.

Back Strengthening

Weak back muscles just can't support your spine and battle gravity for an entire day. If you want your spine to stay healthy and pain-free, your best bet is to strengthen your back muscles with regular exercise.

Ideally, you'll fortify your back with aerobic exercise (walking, running, dancing, and so on) *and* regular strength-training workouts like the ones described in the previous chapter. Core exercises like abdominal crunches (page 85) are particularly helpful because they reinforce the back and abdominal muscles that stabilize and support your spine.

Note If you have a pre-existing back condition or chronic back pain, doing the wrong kind of exercise (or doing the right exercise in the wrong way) could aggravate the problem. If you're in doubt, check with your doctor, and never perform a back exercise that causes any sort of pain.

The single most useful exercise for strengthening your back and relieving back pain is surprisingly simple and remarkably portable. You can do it on the floor of your home gym, in the chair at your office desk, or against the wall at a cocktail party. Pregnant women can use it to alleviate the excruciating backaches caused by their still-in-progress bundle of joy (on the floor up to the fourth month, against a wall thereafter). It goes by many names, including the one used in this chapter: the *pelvic tilt*.

Here's how to do the pelvic tilt:

1. Lie on your back with your knees bent and your hands resting at your sides. (Or stand so you put your whole body—heels, bottom, and shoulders—against a wall.)

2. Tighten your abdominal muscles and squeeze your lower back down against the floor (or wall, or chair). Alternatively, you can rest your hand on the small of your back and push against that.

3. Hold the squeeze for five seconds, but don't hold your breath.

4. Relax your muscles, then repeat this sequence 10 to 15 times.

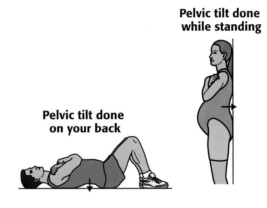

Pelvic tilt done while standing

Pelvic tilt done on your back

Note You can find many of the core exercises cited in this chapter described online. To get started, you'll find good collections of back exercises at *www.mayoclinic.com/health/back-pain/LB00001_D* and *http://tinyurl.com/cyybg9*. (To save yourself some typing, you can find this link on the Missing CD page at *www.missingmanuals.com*.)

5 The Doors of Perception

Perception is a partnership between your sense organs—body parts like your eyes, ears, nose, and tongue—and your brain. Together, they collect clues about the outside world, such as how it looks, sounds, smells, tastes, and feels.

Thanks to the relentless force of evolution, every type of animal has its own carefully tuned senses, and so each lives in a slightly different bubble of perception. For example, although the presence of the earth's magnetic field is obvious to even the daftest migrating bird, your body is completely unable to detect it. Similarly, your senses are closed to the ultraviolet light that guides bees to their favorite flowers and the electric currents that help sharks find their prey. Instead, our senses are tuned to the parts of the world that do the most to further *human* survival. So it should come as no surprise that there are wavelengths of light you'll never see, frequencies of sounds you'll never hear, and chemicals that won't trigger the slightest reaction from your nose or tongue. All your senses are constrained by limits like these. When you look, listen, smell, or taste, you're peering out into the world through the narrow chinks of the cavern you inhabit.

The more you explore your senses, the more you'll realize that they don't just pick up on external reality. Instead, they collaborate with your brain to *convert* a portion of reality into something that's meaningful to you. That's how slightly different wavelengths of light become a range of vivid colors, and how patterns of shifting air pressure become everything from a piano sonata to toddler speech. Behind every human sense is a massive translation effort shaped by your senses and orchestrated by your brain.

In this chapter, you'll explore how your senses make sense of the world. You'll examine your sense organs and see how they work their magic. You'll also learn some practical tips to spot trouble and keep your perceptual hardware in good working order.

Note: In this chapter, you'll focus on how your sense organs detect information and deliver it to your brain. However, a good part of perception is how your brain *processes* this data, which is a shifty subject that you won't explore in this chapter. If you're interested in taking a deeper look into the dark recesses of your brain, check out this book's companion: *Your Brain: The Missing Manual*.

Frequently Asked Question

How Many Senses Do Humans Have?

It's commonly said that the average person has *five* senses. These allow you to see, hear, smell, taste, and touch. Some people add a sixth sense to the mix, which—depending on your perspective—is either a way to talk to dead people or a Hollywood blockbuster with a silly trick ending.

On closer examination, the tidy separation of perception into five senses is a huge simplification of your body's messy biological machinery. What we classify as just one sense is often the result of a number of different mechanisms at work simultaneously. For example, we can divide *touch* into several skin-based sensations, such as the ability to feel pressure, movement, pain, and heat. What we call *smell* includes hundreds of distinctly different types of chemical receptors packed together at the top of your nasal cavity. And, depending on your perspective, what we call *hearing* is actually a type of pressure detection—in other words, your ears detect sounds using a highly specialized form of touch.

Although the scientific basis for counting five senses is shaky, it's still a useful organizing principle for learning about your body. This chapter features separate sections that consider hearing, vision, smell, and taste. To learn about the touchy-feely sensations of your skin, flip back to Chapter 1.

Vision

Your sense of *vision* is a miracle of body perception. With the help of your brain, two fluid-filled spheres convert patterns of light into shapes, colors, and three-dimensional scenes.

Your eyes sit buried in your skull, under flaps of skin that make them look like ovals. If you could remove your eyeballs from their bony, fat-padded sockets, you'd find that they're almost exactly the size and shape of ping-pong balls. Each eye has six muscles strapped to its sides, which pull it in just about any direction, much like puppet strings control a marionette. Though they don't look like much, these muscles are among the fastest and most fatigue-resistant in your body.

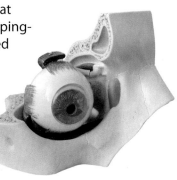

Your Eyes In Motion

Eyeballs are often compared to cameras, but that analogy is a bit off. Your eyes don't record a "snapshot" of what's in front of you—they construct the scene in close collaboration with your brain, sewing together many smaller images to form a cohesive whole.

Here's how it works: Though you may think you're always staring straight ahead, your brain actually keeps your eyes in constant motion, jerking them around with snappy—and subconscious—movements called *saccades*. Thanks to these saccades, your entire visual field is rich in detail, even though your eyeball sees only a dime-sized region of sharpness at a time. Essentially, your eyes collect these tiny fragments of detail as they flit back and forth, and your brain stitches them together into a single, seamless scene.

Saccades are particularly important when you read books like this one. As you read, your eyes jump from word to word, occasionally skipping one (or several, if you're skimming the text). On any given stop, your eyes pick up only a short sequence of letters. This rarely causes a problem, however, because your brain fills in the rest. When scientists use sophisticated equipment to track eye movement during reading, they find a pattern of saccades like the one shown on the next page. (The colored spots indicate where the eye comes to rest; the thin lines show its movement from one spot to the next.)

First text sample:

This person is reading the text carefully in order to understand it. The pattern of fixation shows that the reader is examining most of the words, but not all.

Second text sample:

This person is skimming the text. The pattern of fixations is more dispersed and of shorter duration, which is typical for this type of reading The reader may understand the overall concept, but is likely to remember fewer details.

The Adjustable Eye

To actually *see* anything, you need a way to focus the rays of light that strike your eye down to a near point. Your eye performs this feat using two lenses: the *cornea* (a dome-shaped lens that sits on the surface of your eye) and the confusingly named *lens* (an adjustable lens that lies just inside your eye). Although your cornea does most of the focusing, the lens is arguably more important, because you can flex it. This lets you focus both near and distant objects.

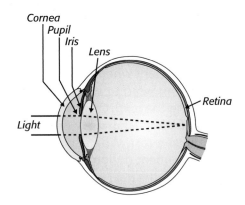

The focused light strikes your *retina*—the lining of photosensitive cells that coats the back of your eye. As light falls here, it excites these photo-receptors, triggering a pattern of electrical activity that's funneled to your brain. From this point on, your neural circuitry does most of the heavy lifting, converting colors, edges, and patches of light and dark into shapes and three-dimensional objects.

The lens isn't the only adjustable part of your eye. Your *iris*—the colored part of your eye—expands or contracts to change the size of your *pupil*, determining the amount of light it lets into your eye. (Incidentally, your iris can't close completely, which is why it's a bad idea to stare into the sun. Even when fully contracted, your iris may let in enough light to damage your retina.)

Note You may have heard that your eye creates an upside-down image on your retina. This is true. You may also have heard that your brain is responsible for turning that image right-side up. This is nonsense. (To understand why, imagine turning over a video recorder. Will it record television shows upside-down?) From your brain's point of view, it makes no difference whether an image is rightward, backward, upside down, or inside out, as long as your brain has the wiring necessary to *analyze* that image.

Your Faulty Focus System

Although your eyes have an impressive focusing system, it doesn't always live up to its potential. In fact, the odds are that your eyes—like those of most people in modern society—have some sort of refractive shortcom-ing. The problem is usually that the shape of your eye and cornea is dis-torted, so that no matter how your lens adjusts itself, it can't project a clear image onto your retina.

Here are the most common optical problems:

- **Nearsightedness.** Your eye can't focus on distant objects. The cornea may be overly curved.

- **Farsightedness.** Your eye can't focus on close objects. The cornea may be overly flat.

- **Astigmatism.** The surface of your cornea is irregularly warped, making objects both near and far appear blurry.

The pictures on this page show how these refractive problems scramble your vision. To see clearly, your eyes must focus nearly parallel rays of light so they converge on your retina at the back of your eye. If the light doesn't converge in the right spot, you perceive a fuzzy image.

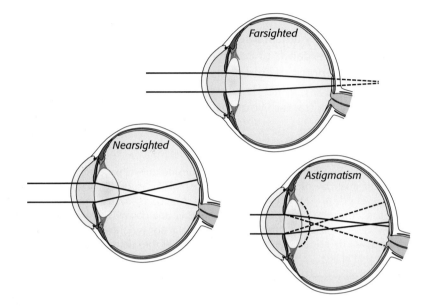

The simplest way to check your vision is with the well-known Snellen chart, that series of shrinking letters that adorns your eye doctor's wall. (To find a version you can print up and use on your own, see *www.i-see.org/eyecharts.html*.) Here's a tip to get you started: the first letter is nearly always *E*.

Usually, you stand 20 feet from the Snellen chart. If you can read the 20/20 line of the chart from this distance, you have 20/20 vision. That's the gold standard for humans. In comparison, a nearsighted person with 20/40 vision can just barely make out at 20 feet what a normal person can see from 40 feet. (Of course, a "normal" person is defined as someone with 20/20 vision, even though most people have noticeably poorer vision, as described in the box on the next page.)

Some people have better than 20/20 vision, and many people have a mixed pair of eyes, with one stronger than the other. In the U.S., legal blindness is defined as 20/200 vision or worse.

Design Quirks

Flawed Vision

Our eyes have a clear and obvious design flaw—they often fail to focus properly. This is a far greater problem than the eye's other quirks, such as its blind spot (page 115) and its inability to see certain colors (page 114). So why should such an elegant, intricate organ have such a glaring hitch?

Although there's no definitive answer, many experts now believe that the problem has something to do with modern life. In the distant past, our ancestors spent most of their time wandering the savanna and staring into the distance. In today's world, we spend far more time on *near work*—close-up activities like reading, writing, and drawing. During childhood, these activities may gradually reorient the eye from its natural state of mild farsightedness to the far more common condition of adult nearsightedness. In fact, our vision may have worked far better in the era before the invention of corrective eyewear (which was developed relatively recently, in 13th-century Italy).

Although this is all somewhat speculative, it may eventually lead to a way to prevent eye trouble. For example, children might be able to perform far-vision exercises to counterbalance the overload of near work. But for now, there's no solution other than glasses and contact lenses, and there's no evidence to suggest that you can help your eyes by cutting down on modern activities in adulthood. So now that your vision is already screwy, it's safe to keep reading, writing, watching television, and surfing the Web.

No matter how good your vision today, as you age, your eye's lens stiffens and its focusing muscles weaken. The result is *presbyopia*, which is a very intimidating way of saying that you'll have trouble focusing on close objects.

In fact, the changes that underlie presbyopia begin very early in life. Small children can focus on objects extremely close to their eyes, while adults need to step farther and farther back. Eventually, holding objects even at arm's length isn't far enough away to focus clearly, which is why this condition is sometimes called "short-arm syndrome" (as in "My arm's too short to read the writing on this blasted prescription bottle!").

Most people notice their reduced focusing power after age 40. One popular solution is reading glasses (which tend to migrate away from any important reading material once you put them down). If you already wear corrective glasses, you can switch to bifocals or progressive lenses, which are essentially two ways of combining reading glasses with your existing prescription. If you wear contact lenses, you can try *monovision*, which is a more radical solution that corrects one eye to see distant objects while magnifying close objects for the other eye. Both solutions require a period of adjustment before your brain figures out how to take advantage of its new optical hardware.

Tip One thing that *won't* damage your eyes is reading under dim lights. Yes, the practice can cause strained eyes and a headache (much as your mother warned), but the effects are temporary. To really protect your eyes, you should stop worrying about the dark and start fearing the *light*. Over time, the harsh ultraviolet rays of the sun on a summer day can contribute to eye diseases like *cataracts* (which cloud your lens) and *macular degeneration* (which causes patches of blurred vision and blindness). The best protection against problems like these is to slap on a pair of UV-protective sunglasses.

Seeing in Color (and at Night)

As you've already learned, your retina is packed with light-sensing cells. But you may not realize that these light detectors come in two flavors:

- **Rods.** These are the most numerous light-sensing cells in your eyes. They're extremely efficient at collecting light and continue working even in near-darkness. For this reason, your rods power your night vision. However, rods don't distinguish between different colors.

- **Cones.** These cells are less sensitive than rods (meaning they need more light to do their work), but they're able to perceive fine detail and color. Your eyes have three types of cones, which give you the ability to perceive three primary colors (red, green, and blue), provided there's enough light for your cones to function.

Rods are spread throughout the sides of your eye. Cones are packed into a small region in the middle of your eye. If you created a map of your eye, it would look something like this:

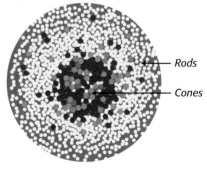

Rods

Cones

This combination of rods and cones gives your eye the best of both worlds—highly sensitive black-and-white vision you can use to creep around at night, and sharp color vision that catches more detail during the day. Although your brain switches effortlessly between the two, it takes 30 minutes of darkness to develop your best night vision. That's because night vision requires chemical changes in your eye that make your rods more sensitive.

A short blast of ordinary light usually resets your vision from dark- to light-adjusted. However, dim red light doesn't trigger this change because your dark-adjusted rods are naturally tuned to the blue end of the light spectrum, and they have a hard time picking up the color red. That's why astronomical observatories and submarines use red light to illuminate dials and switches—it provides enough light for operators to see their instruments in a darkened environment, but it doesn't disrupt their night vision.

Incidentally, most color-blind people have the same three types of cones as everyone else, but their cones don't function normally. These people may have trouble discriminating between certain shades of color, but the effect is usually minor. Often, it goes unnoticed until they try a color-vision test (like the spot-the-number test at *http://en.wikipedia.org/wiki/Ishihara_color_test*). However, people with rarer, more serious forms of color-blindness may have only two functioning types of cones. They may be unable to distinguish purple from blue or red from yellow, except by the slight changes in brightness of the hues, not by the colors themselves.

Binocular Vision

Having two eyes (which biology nerds call *binocular vision*) gives you a slightly wider field of vision than you'd get with just one eye. But the real advantage of binocular vision is that each eye sits in a slightly different position in your skull, giving it a slightly different view of the world. This difference gives your brain the clues it needs to make accurate predictions about distance—for example, to tell you how far to lean forward to grab your coffee cup off the table.

We're All Color Blind

Black-and-white vision is a one-dimensional affair. A given shade can get blacker or whiter, but that's it.

Color vision (by which we mean *human* color vision) has three variables. You can modify any color by changing the amount of red, green, or blue in it. This mix of colors triggers different cones in your eye, creating the perceptual experience of seeing a single, specific shade.

But some animals aren't limited to three types of cones. Consider birds, whose eyes have a number of advantages over yours. Their eyes are stacked with more cone, and these cones are arranged in larger patches (or in multiple patches) and packed more closely together. This arrangement gives some birds spectacularly sharp vision over long distances. But the most remarkable feature of birds' eyes is how their cones *work*. Unlike your eyes, which have three types of cones, birds' eyes have *four* or *five* distinct types of cones.

So what does this mean? Imagine meeting a pigeon and traveling with it to the countryside. Your eyes will collect the same reflected light as the pigeon's eyes. You'll see the same scenery. But the pigeon will perceive that light differently. Its eyes will break the rolling hills into a mix of four primary colors, while you translate them to a measly combination of three. For the pigeon, the contrast of certain wavelengths of light will become more dramatic, allowing it to spot details that your eyes miss. The difference is a little like leaping from black-and-white to color vision. Ultimately—and there's no way around this—the pigeon will get a subtler, more nuanced view of the outside world.

So the next time an interior decorator scolds you for mistaking pistachio green for chartreuse, remind yourself that in the eyes of a pigeon, we're all color-blind.

Many other animals lack refined depth perception, but compensate with better peripheral vision. For example, most birds and lizards have eyes on either side of their skulls. By pivoting each eye separately, they can see nearly 360 degrees—the entire way around their bodies. For what it's worth, *prey* animals—those that need to spot trouble quickly and run— are more likely to have wraparound vision than *predator* animals, which are more likely to have binocular vision.

Binocular vision also means that you never notice your *blind spots*—two sizeable holes in your field of view (one in each eye) where nothing appears. These blind spots occur where the optic nerve exits your eyeball, but your brain fills in those spots by combining the information it collects from both eyes.

To find your blind spot, start by closing one eye so your brain can't perform its fill-in magic. Even now, you won't notice the blind spot, because your brain uses saccades (page 107) to gather information and stitch together a complete scene. But if you lock your eye in place and put something conspicuous directly into your blind spot, you'll see that "something" disappear.

To try this out, hold this book very close to your face, close your left eye, and look at the X in the picture below with your right eye. Now slowly move the book away from your face, until the O disappears into your blind spot.

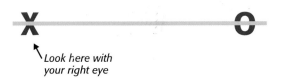

Look here with your right eye

Interestingly, even when the O disappears, you still see an unbroken yellow line. That's because your brain fills the line in automatically, making its best guess about what probably exists in the spot it can't see.

> **Note** This blind-spot test hints at one of the more enjoyable parts of perception: optical illusions. Although you might assume that optical illusions are rooted in the behavior of the human eye, most of them exploit the quirks of your brain as it converts contrast and color into shapes and moving objects. You can find more about perceptual illusions in *Your Brain: The Missing Manual*.

Hearing

Your sense of *hearing* is one of those things that seems straightforward at first, but becomes increasingly bizarre the more you dig into it.

The basic principle is simple: Certain events—a book tumbling off a shelf, a twig snapping in the forest undergrowth, two cymbals clashing together—disturb the air around an object. That object (the book, the twig, the cymbals) vibrates, and these vibrations knock around the nearby molecules of air. Turn the page to see how it works with a drum.

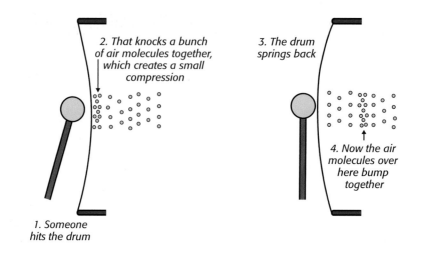

2. That knocks a bunch of air molecules together, which creates a small compression

3. The drum springs back

4. Now the air molecules over here bump together

1. Someone hits the drum

To make a perceivable sound, an object needs to vibrate many times each second. Each vibration knocks some air molecules out of place, and these air molecules bump into more air molecules, which smack into still more air molecules. This creates ripples of air pressure that spread out in all directions, and we call these ripples *sound waves*. The violence with which the air molecules bounce back and forth determines how *loud* the sound is, and the number of vibrations that take place each second determines its *pitch* (whether it sounds like a low hum or a piercing, Pavarotti high C).

So sound is really nothing more than tiny fluctuations in air pressure. From this perspective, there's no good or bad sound, no tuneful or discordant noise, and no clear difference between gangsta rap and your neighbor's car alarm (which is something you may have already suspected).

Converting Pressure Puffs to Sound

Somewhere along the line, in the billions of years of evolution of life on earth, life decided that it would be worthwhile to pay attention to the air vibrations we call sound. Different species used them to detect potential threats (like prowling cheetahs), to find food sources (whatever's hiding from the cheetah), and even to communicate ("Hey, look! There's a cheetah!"). Shortly after humans developed the ability to hear, we invented *ignoring*, mostly as a way to facilitate marital negotiations involving dirty dishes, soiled laundry, and late-night poker games.

Exploring Your Frequency Response Curve

An average young person can hear a staggering range of sound vibrations from 20 Hz to 20,000 Hz. (A Hertz is one vibration per second. So 20,000 Hz is a vibration that happens 20,000 times each second.) Of course, you won't hear all pitches equally well. The middle-range pitches are the easiest to detect. A sound that's similarly loud but extremely high-pitched will seem much quieter to your ear. You'll notice a similar fall-off as you descend to the rumbles of low-frequency sound, although you'll begin to *feel* them reverberate through your body.

As you age, the upper range of your hearing shrinks. For example, the hearing of a normal, middle-aged adult tops out at a significantly lower (but still impressive) 14,000 Hz. This change has been used to some effect by clever people on both sides of the age divide. For example, one company sells a sound device called the Mosquito Ultrasonic Teen Repellent (*www.noloitering.ca*). It emits an annoying sound that only young people can hear. Put it in a place where you don't want crowds of teenagers to loiter—say, a mall parking lot—and it's sure to drive them away. A similarly inventive product is the Teen Buzz ringtone (*www.teenbuzz.org*). Teenagers use it to announce text messages on their cell phones without alerting teachers and other authority figures in the over-30 crowd.

To test your own hearing and find the highest frequency you can detect, try an online hearing test (like the one on *www.phys.unsw.edu.au/jw/hearing.html*). Of course, these tests depend on some factors you can't control, such as the quality of your speakers and the noise your computer makes, so they're not perfectly accurate.

From a scientific perspective, this all makes perfect sense. But here's the strange part—you interpret sound in a way that's distinctly different from the way you interpret other sensory information. When a warm breeze blows against your face, you have the distinct impression of, well, a warm breeze blowing against your face. In a similar way, you might expect to feel sound as a series of little air puffs against your skin. But what you actually perceive is a rich range of tones, from rumbles to whistles. And if the right tones occur in the right patterns, you don't perceive distinct air vibrations at all—instead, you hear music and speech.

The organ responsible for this remarkable transformation is your *ear*. The following sections explain how it works.

The Outer Ear

Your body's sound-collection system starts with your *outer ear*. The most obvious parts of your outer ears are the two flaps of skin-covered cartilage that adorn either side of your head. They're called *auricles*, and while they usually get the most attention, they're surprisingly inessential. Their only purpose is to direct sound into your ear canal, somewhat like a funnel. In addition, their design may help prevent unwanted substances (flies, dirt, chicken wings, and so on) from lodging themselves inside.

Outer ear

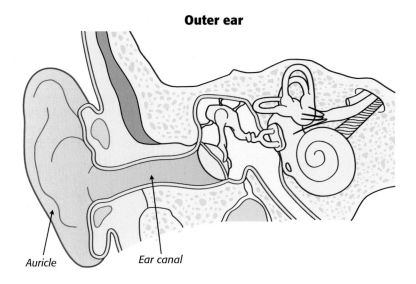

Auricle   Ear canal

Some animals can move their auricles to better capture the sounds they want to hear, but most humans can't. If someone were to lop off your auricles, you might have trouble keeping your sunglasses on, but you'd still be able to hear perfectly well. (In fact, many animals—like snakes—lack external ears altogether, but they can still hear. Sound waves travel through their skin and muscles, and then vibrate the ear bones inside their head.)

The auricle leads to the *ear canal*, which is also considered part of your outer ear. The ear canal is a simple tube that leads sound down to your eardrum. Its real purpose is protection—it uses thick, sticky earwax to immobilize dirt, bacteria, and insects, much like a strip of bug-catching tape.

Removing Earwax

On a good day, earwax does its job, trapping foreign material and gradually moving up your ear canal to the opening of your ear. There, it dries up and eventually falls out, its job done.

But earwax isn't always so accommodating. Many people have particularly hard or sticky earwax. It becomes trapped inside the ear canal and can eventually plug it up, causing pain and blocking sound. In fact, this problem can happen to anyone, because the consistency of earwax can change abruptly and without explanation.

So what should you do about earwax? Don't start digging with a cotton swab. Contrary to what you might have heard, they *are* safe for removing earwax, but only if you limit your swabbing to the opening of your ear canal and resist the urge to plunge in. Deeper swabbing is likely to compact wax, turning a partial blockage into a complete one. It's also dangerous, because it risks damaging the sensitive eardrum or scratching the ear-canal skin, which can lead to a painful infection.

So what can you do? If you don't have problem earwax, you don't need to do anything—after all, the wax in your ear canal belongs there, and it will migrate out in its own good time. But if you're prone to wax of the ear-clogging variety, a simple practice can help prevent blockages. Put two or three drops of mineral oil into each ear every day (using an eye dropper). Over a couple of weeks, this softens wax so that you can remove it with a gentle flush of water. If your ear is completely blocked, head to your family doctor, who can wash it out or scoop it out using special tools.

The Middle Ear

The middle ear is where the waves of changing air pressure start their transformation into sound. First, the waves strike your eardrum, which sits at the end of your ear canal, making it vibrate like a speaker cone. These vibrations then pass through a set of three bones called *ossicles* (which, incidentally, are the smallest bones in your body). Because the ossicles transfer sound vibrations from your eardrum to a much smaller surface, they amplify the sound, making it at least 22 times stronger. Without these tiny bones, you'd miss out on rustles, whispers, and just about everything fainter than an industrial vacuum cleaner.

Middle ear

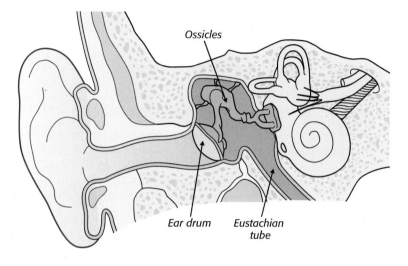

The middle ear is lined with mucus and filled with air. This causes trouble when you experience a sudden change in air pressure (when you go up or down in an airplane, for example). In such a case, the pressure difference between the inside and outside of your ear pushes or pulls your eardrum and causes pain. You relieve this pain by equalizing the pressure, using a passageway called the *Eustachian tube*, which runs from your inner ear to the back of your throat. It's usually pinched shut, but chewing, swallowing, or yawning opens the tube, relieving the pressure.

> **Note** Ordinarily, the Eustachian tube helps drain mucus out of your middle ear. But it's also a potential access route for bacteria, and the source of painful ear infections. The problem is particularly common in children, because their Eustachian tubes are shorter and more horizontal.

The Inner Ear

The inner ear is home to two very different senses.

First, there are the *semicircular canals*—three tubes that control your body's sense of balance. These tubes are positioned at right angles to each other, allowing them to perceive motion in all directions—up, down, left, right, forward, and backward. As you move around, fluid swishes through these tubes, bending the microscopic hairs inside and allowing your brain to determine how your body is moving through space. Spin around long enough, and you'll temporarily overwhelm this system, which delights small children and gives whirling dervishes a transcendental experience (*http://tinyurl.com/6r9z6c*).

Second, there's the snail-shaped *cochlea*, which is the final stop in your body's sound-processing system. Like the semicircular canals, the cochlea is filled with fluid and lined with hairs, but the similarities end there. When the bones in your middle ear strike the cochlea, its fluid begins swishing back and forth, tugging on the thousands of tiny hairs inside. Different pitches cause the fluid to swish in different ways and to vibrate different hairs. The result is a pattern of electrical stimulation that's sent to your brain and transformed into sound. Incidentally, these hairs are also the most common source of hearing loss. If they're damaged or destroyed, your body can't replace them.

Inner ear

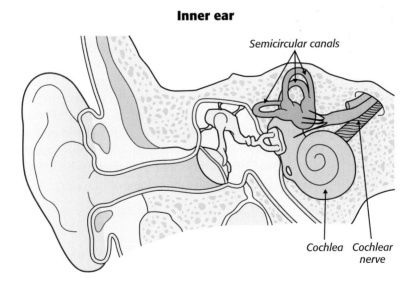

Semicircular canals

Cochlea   Cochlear nerve

The more you learn about the cochlea, the more you realize that your sense of hearing is full of quirks. Here are some key examples:

- **Uneven pitch detection.** The microscopic hairs in the cochlea are spread out unevenly, giving you vastly better hearing at certain frequencies—for example, at the frequencies of human speech.

- **Noise cancelling.** The cochlea works with your brain to identify background noise and filter it out, sort of like a pair of high-tech, noise-cancelling headphones. That's why it's fairly easy to focus on the conversation that interests you at a cocktail party, but maddeningly difficult to get the same information if you listen to an audio recording of the party—you can't perform the same noise-filtering feat after the fact.

- **Masking.** Masking refers to a number of subtle phenomena that prevent you from hearing certain auditory details. For example, if you listen to loud music, your ear adapts to the heightened volume and might miss quiet sounds that it would ordinarily hear perfectly well. This effect is stronger when the overlapping sounds have similar frequencies, and it's always easier for low-pitched sounds to drown out high-pitched ones. Another type of masking occurs when a sound ends. It takes a short but measurable period of time for your sound receptors to recover, and during this time you may be unable to hear a similarly pitched sound.

These fine details might seem like little more than fun trivia, but they actually shape everything we do with sound. It's no exaggeration to say that the structure of the ear determines the structure of a classical symphony, influencing everything from how we build instruments (making sure they produce the loudest, richest tones in the range we hear best) to how we mix high-pitched and low-pitched sounds (making sure we don't drown out the tones we want to hear). Another example is audio-encoding technology—for example, the kind that shrinks an avalanche of digital audio data into a lightweight MP3 file. To perform this sleight of hand, audio encoders toss out a great deal of information that they decide you won't hear (most of the time), based on the complex principles of masking.

Understanding Volume

Your ear takes in a lot of information. Just as it can stretch itself to accommodate a wide range of frequencies, it's also able to deal with a staggering range of volume—everything from a whisper to a train whistle (which is 10 million times as intense).

Note To adapt to different volume levels, your ear uses a miniscule muscle that dampens the movement of the tiny ear bones that convey sound. Your ear is so paranoid about hearing loss that it contracts this muscle the moment you open your mouth to speak. However, your ear can't anticipate sudden volume changes, which is why a gunshot damages your hearing more quickly than the continuous racket of a pumped-up iPod.

People measure sound intensity using the *decibel* scale, which starts at 0 (representing a barely audible sound). Each increase of 10 decibels represents a sound that's 10 *times* more intense. For example, a 30-dB sound is three 10-dB increments from 0, which means it's 1,000 times ($10 \times 10 \times 10$) more intense.

However, this isn't the way we *perceive* sound. To our remarkably flexible ears, a sound that's 10 times more intense seems merely twice as loud. Similarly, the difference between 0 dB and 30 dB feels like a tripling of volume. That's part of the reason that some people can gradually and unwittingly get used to eardrum-popping volume levels.

Incredibly, your hearing range spans from 0 dB to 120 dB, after which increased volume will cause significant pain. Here are a few typical volume levels:

Sound	Volume
A barely audible pin-drop	0 dB
A whisper	15 dB
A normal conversation	60 dB
A noisy restaurant	70 dB
An average lawnmower	90 dB
A car horn	110 dB
A rock concert	115 dB
An iPod at maximum volume	100–120 dB
A gunshot	140 dB

Your ear requires very little care. To keep it in good working order, avoid prolonged exposure to any noise of 85 dB or above. That means wearing earplugs in noisy environments (if you work in a machine shop or perform with a rock band, for example). Similarly, exercise caution with portable music players. If you rock out at high volume for eight hours a day, you're boogying your way to early-onset hearing loss. And finally, pay attention to what your ears tell you. A faint, persistent ringing (called *tinnitus*) is a common sign of temporary hearing damage after short-term exposure to loud sounds (say, during last night's club crawl). If you tune out that warning, you're on the path to permanent silence.

> **Tip** As a rule of thumb, if you need to raise your voice to drown out a competing sound, you're listening to something in the 85-dB range.

Smell

Smell is one of two senses that's all about detecting chemicals (the other is taste). When you smell something, you detect chemicals wafting through the air. When you taste something, you detect chemicals you've put in your mouth.

You may think you smell with your nose, but science has shown that this makes about as much sense as thinking you hear with your earlobes. In fact, your nose is simply a glorified intake and exhaust pipe. It pulls air in, directs it through your *nasal cavity*, and sends it down to your lungs. The nasal cavity is where all the smelling action takes place.

Into the Nasal Cavern

Your nasal cavity is an unpleasant place. It's a vast, mucus-coated cavern that conditions air—warming it up and filling it with moisture—before it reaches your lungs. As air streams through your nasal cavity, it drifts past two very small, dense patches of tissue at the top. Each patch is about the size of a postage stamp and is packed with millions of exquisitely sensitive *odor receptors*. Each receptor is designed to identify a particular type of chemical.

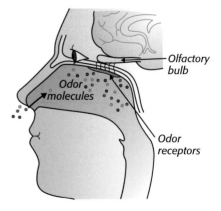

Olfactory bulb

Odor molecules

Odor receptors

To smell something, an airborne molecule must first dissolve in the mucus that lines the top of the nasal cavity. It sinks through this thick layer and settles onto one of the microscopic hairs of an odor receptor. This triggers an electrical signal that travels to the smell-perceiving part of your brain—the *olfactory bulb*.

You can't smell every type of chemical. The chemicals you *can* detect are called *odorant molecules*, and they all have a few things in common. First, they're light, which lets them vaporize easily and drift away from the object you smell, then float through the air and into your nose. Furthermore, odorant molecules dissolve in both water and fat, which lets them penetrate your nasal mucus and bind to an odor receptor.

> **Note** It may seem a bit odd, but you can't smell anything without pulling part of it up into your nose. This isn't much of a concern if you're enjoying the aroma of freshly squeezed orange juice. However, it's a bit more disquieting when you find yourself at the beach on a summer day, in the men's washroom, and the toilets are clogged.

Rough estimates suggest that you have about 1,000 different types of odor receptors in your nasal cavity. However, it's only when odorant molecules trigger odor receptors in the right *combination* that the message passes to your olfactory nerve and your brain perceives a smell. That's why current thinking suggests that humans can recognize a staggering 10,000 different aromas. (This is a bit like the way color vision works. Every shade of color you see is built out of a combination of just three primary colors [red, green, and blue] that your three types of cone cells can detect. But in your nose, every smell is built out of a combination of the *hundreds* of different types of odor receptors. This gives your sense of smell a much broader range than your sense of vision, and lets you detect everything from toasted cumin seeds to sweaty gym socks.)

Interestingly, odor perception varies from person to person, and some people are "blind" to certain smells. To evaluate your sense of smell, you need a doctor-administered scratch-and-sniff test.

> **Note** Unfortunately, because there are so many different smells and so little that relates them to one another, it's hard to describe what you're sniffing. Although we have a rich vocabulary for describing color, relatively few words concretely describe different aromas. Most of the time, all we can do is point out what causes the smell (for example, "smells like bacon" or "smells like burning rubber").

The Purpose of Smell

Although the sense of smell gets little respect, it's a surprisingly practical tool for discovering the world around you. In fact, all living creatures have some sort of chemical-detection mechanism, which means that smell is probably the oldest type of perception known to life.

As you sniff around, you'll notice that some odors elicit a nearly automatic response. These smells help steer you away from danger—like rotting fruit you definitely shouldn't eat, or festering feces you definitely shouldn't touch. But most odors depend on powerful *learned associations*. For example, you assume that a house is clean when your nose detects smells like bleach, ammonia, and the fragrances of cleaning products. Although these smells might make you feel comfortable enough to stay for dinner, there's no guarantee that the kitchen countertops aren't colonized with the very worst strains of stomach-churning bacteria. The smell of "clean" is simply a learned clue from which you make a reasonable assumption—an assumption that may or may not hold true.

> **Note** Given the close tie between odor and memory, it's no surprise that a faint whiff of the right scent can bring back a vivid, long-forgotten memory. This phenomenon is called *Proustian memories*, after the way the smell of a cookie kicks off Marcel Proust's seven-volume epic of reflection. (Less appealingly, the dank smell of a public lavatory fuels an equally vivid series of memories later in the story.)

Smell often functions as an early warning system. For example, a burning smell while you prepare Sunday dinner quickly puts you on high alert. Smells also help you find and recognize the things you want in life—whether that's the aroma of apple pie in the oven or the familiar scent of a romantic partner in your arms. (In fact, we often underestimate our ability to recognize people by body odor. It starts off as an essential survival skill for infants, but remains active throughout our lives.

People who develop *anosmia*—a medical condition where the sense of smell disappears entirely—suffer from a range of problems. They may find eating a tiresome chore, or they may eat incessantly in a futile attempt to taste *something*. They may miss important warning signals (like spoiled food or a dead cat under the porch). Finally, people who develop anosmia as adults often suffer from depression and a reduced sex drive.

Odor Fatigue

Your brain quickly adapts to new smells. What smells pungently strong on the first sniff becomes almost undetectable a few minutes later. This phenomenon, called *odor fatigue*, is similar to the way the hum of an air conditioner fades into background noise, but odor fatigue is more powerful. That's because once you lose track of a smell, you can't will yourself to get it back—at least not without first leaving the room and smelling something else.

Odor fatigue makes perfect sense when you consider the evolutionary role of smell. Your body is more interested in using smell to *detect* things than it is in keeping track of them. Once you notice a smell and decide how you want to react, it's time to move on so you can detect the next potentially important odor.

Now that you understand how odor fatigue lets you ignore prolonged smells, there are a few practical bits of advice to consider:

- To determine if an object or room in your house smells, start by walking out the door and giving your nose a break. Then step back inside and pay attention.
- To determine if you have body-odor issues, you'll need the help of a friend who can sniff you out.
- Just because you *can* ignore a smell doesn't mean you *should*. Irritants and even toxins, ranging from cleaning products to cigarette smoke, are easy to ignore if you live with them, but they might not be so easy on your lungs. If in doubt, open a window and get some fresh air.

Taste

Your sense of *taste* works much the same way as your sense of smell. The most obvious difference is the location. The chemicals you detect by smell must settle into the mucus at the top of your nasal cavity, but the chemicals you detect by taste must dissolve into the saliva that coats your tongue.

Many people assume that the bumps that cover their tongues are taste buds. They're actually *papillae*—tiny lumps of tissue that help your tongue grip food and move it around while you chew. Most of your taste buds line the grooves around these lumps, but some are scattered along the surface of your mouth and throat.

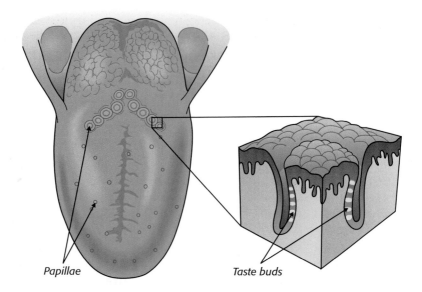

Papillae   Taste buds

Taste Buds

The vocabulary of taste is smaller and simpler than the language of odor. Although scientists aren't in complete agreement, it seems that there are just a few types of taste receptors in a taste bud. While the nose impresses with roughly 1,000 types of odor receptors, there are just five or six types of taste receptors. They detect sweetness, saltiness, sourness, bitterness, and "umami" (often described as the savory flavor of cheese and many Asian dishes). Scientists disagree about whether we have specific taste receptors for fat. For years, they assumed that fat simply improved the texture (or "mouth feel") of food, or that it was responsible for dissolving and delivering other flavor chemicals to your taste buds. But recent studies suggest that your tongue can pick up on fat—and that it likes what it tastes.

> **Note** You may have seen tongue "maps" that show clusters of different taste-bud types in different spots on the tongue. More recent explorations suggest that the tongue is fairly uniform—it has a mix of taste buds spread throughout, with a higher concentration at the sides and tip.

The tongue is an inhospitable place. Grinding friction, searing heat, and potent acids make short work of many a taste bud. Fortunately, your tongue produces replacements, and the average taste bud lasts only a week or two before a successor takes it place. As you age, this replacement process becomes less effective. Although you begin life with about 10,000 taste buds (each packed with up to 100 taste receptors), your count shrinks by half as you enter old age.

Why Do You Crave Sugar, Salt, and Fat?

Your tongue has a limited variety of taste receptors because sugar, salt, and fat were the flavors most important for human survival over the last few million years of evolution. Your tongue craves sweetness because it signals ripe fruit. You yearn for salt because it's an essential compound for basic body function. You long for fat because it's an extremely dense source of dietary energy. On the other side of things, sourness can alert you to spoiled food, and bitterness can warn you about poisonous substances.

Our sense of taste is a great tool when we need to select nutritious, non-toxic foods from a natural environment. However, its effects aren't as positive when we use it to guide food *creation*—for example, when we engineer heavily refined foods like candy-coated breakfast cereal. In this situation, our natural preference for sweet and salty runs rampant, producing foods that are literally too much of a good thing. These foods still taste good on our tongues, but over time they can throw our bodies seriously off kilter.

The good news is that learned associations can gradually trump our built-in drive for sugar, salt, and fat. After all, many highly prized tastes involve a complex assortment of flavors along with the sour or bitter notes we normally avoid. For example, chocolate, coffee, beer, citrus peel, and greens like escarole all have strong bitter notes that we enjoy when matched with other flavors. There's a good evolutionary reason for this flexibility—generations ago, humans who could discover new, untapped sources of food had a huge survival advantage.

This is particularly significant if you're a parent and you want to expand your kid's taste universe beyond chicken nuggets and plain pasta. The best advice is to present new foods, several times, with no conditions. Avoid resorting to bribery, bargaining, or threats, all of which place a taste "value" on food. (For example, using chocolate as a reward for eating spinach teaches that chocolate is desirable and spinach is not.) While you're unlikely to find a child who prefers rapini to peanut butter, in time we can all learn to love a wide range of flavors.

Supertasters

You eat a varied diet and plenty of vegetables. Your best friend subsists on diet soda and cheese pizza. Is she hopelessly spoiled, or is her perception of taste just wired differently? Modern science might have the answer.

In recent years, scientists have begun discussing *supertasters*—individuals who experience taste more strongly than others, and who are more likely to be picky eaters. Supertasters are particularly sensitive to bitter flavors (which they often shun). Research suggests that in the U.S., roughly 35 percent of women and 15 percent of men are supertasters.

If all this sounds made up, you'll be surprised to hear that a scientific test can sift out supertasters. The test is remarkably simple. Participants taste a solution that includes a chemical like *propylthiouracil*. Many people find this taste mildly unpleasant, but supertasters find it stomach-churningly awful. Still other people (called, rather sadly, *nontasters*) say the mystery chemical has no taste at all.

You can test your own tasting abilities using an at-home kit, like the kind sold on *http://supertastertest.com*. Or you can try the crude experiment discussed next, which uses dye to help you estimate the number of taste buds on your tongue. (The current thinking is that supertasters have a heavier concentration of taste buds on their tongues, but this is probably not the only reason for supertasting.)

Here's how to perform the test:

1. Gather up your supplies: some blue food coloring, a piece of paper with a 7-millimeter (about a quarter-inch) hole punched through it, and a magnifying glass.

2. Using a cotton swab, rub some of the food coloring onto the tip of your tongue. Your tongue will absorb the dye, but the tiny papillae (page 127) will stay pink. This is where your taste buds are.

3. Put the piece of paper over the front part of your tongue.

4. Using the magnifying glass, look at the hole. Now count how many pink dots you see within the hole. A score of fewer than 15 suggests you're a nontaster. A count of from 15 to 35 ranks you as normal, while anything greater suggests supertasting abilities.

> **Note** Supertasters don't necessarily have a *better* sense of taste than normal people. It all depends on your perspective. Yes, supertasters have a stronger reaction to certain tastes, but there's still much to be said for a *cultured* palate—in other words, a broad love of food honed through years of comparative tasting.

The Riddle of Flavor

Seeing that your tongue detects only a few different flavor types, it seems odd that we can distinguish between a roast beef and a rutabaga (never mind between different vintages of Zinfandel).

A good part of this disparity is thanks to your nose's chemical-sensing abilities. As you eat, volatile chemicals stream off the food in your mouth, rise through the back of your throat, enter your nasal cavity, and arrive at the same odor detectors you saw on page 124. These detectors work in conjunction with the primary tasting ability of your tongue to generate endless variations of flavor. In fact, some studies suggest that the tongue is a bit player in the sensation of taste, and that 80 percent of taste takes place in your nasal cavity.

Note A heavy cold can clog your nose and fill your nasal cavity with extra mucus, making it all but impossible for chemicals to reach your odor receptors. The result is a flattened, blander sense of taste.

For many years, food scientists thought the interplay between tongue and nose was the whole story behind taste. However, more recent research shows that taste involves an impossibly complex mix of factors:

- **Texture.** The feel of food as it gives way to your teeth and mixes in your mouth changes how you feel about it. For example, study participants rank a thick cheese sauce as tasting cheesier, even if extra flour is the only difference. And if you doubt the power of texture, try putting some hot, crispy french fries in a plastic bag for a few minutes, then see if you feel the same way about the identically flavored soggy sticks that emerge. Or compare the sweetness of a fizzy drink with an identical one that went flat half an hour ago.

- **Temperature.** Hot food is tastier because its volatile chemicals are more likely to drift to your nasal cavity and trigger different odor receptors. However, even without this factor, the temperature of food influences your overall experience. Think, for example, how soup warms your stomach, and why you'll never mistake a spoonful of ice cream for butter or yogurt.

- **Touch (or pain).** Some sensations that happen on the tongue have nothing to do with your taste buds. Examples include the tingle of alcohol, the coolness of mint, and the sear of hot spices. These compounds mess with the other nerve endings in your tongue, and some, like hot chile peppers, actually cause pain. (Chile-pepper survival advice: Because the active ingredient is an oil, water does little to wash it away. The fat in whole milk or the alcolhol in beer dissolve it more easily. By comparison, the singe of mustard and wasabi is shorter-lived because most of its kick comes from the odor receptors in the nasal cavity.)

> **Note** Strong spices likely evolved as a way to kill bacteria and disguise the off-flavors of rancid food. That's why mouth-scorching cuisine appeared first in tropical climates, where food spoilage is a greater concern.

- **Appearance.** Cooks are taught that the first bite is with the eye, and there's a solid body of science that suggests they're right. Just changing the color of foods causes people to imagine different tastes. For example, study participants will stubbornly describe orange-flavored, yellow-colored Jell-O as tasting like pineapples. A similar effect has been noticed with white wine that's dyed red.

- **Combinations.** When you pair different flavors, the result is more than the sum of its parts. For example, tasters may claim a strawberry tastes more strongly like strawberries when it's diluted with water and sweetened with sugar.

- **Expectations.** There's a reason that food tastes better when you dine at a fine restaurant. The framework of assumptions and expectations you have when you approach a meal conditions your brain to perceive it in a certain way. Current emotions and past experiences are particularly powerful: from fond childhood memories of Grandma's apple pie to the time you projectile vomited mint-chocolate ice cream on the Tilt-A-Whirl.

In short, your perception of flavor relies on *all* your senses, your memories, and your current state of mind. It's one of your body's great surprises: A simple sense that seems to involve little more than a few bumps on your tongue just might be your body's strangest door of perception.

6 Your Lungs

nhale. Exhale. Repeat 30,000 times.

That doesn't make for a riveting diary entry. But from your lungs' viewpoint, every day is pretty much the same. You go about your daily life. Your lungs process lungful after lungful of air. The result? Whether you need to climb a flight of stairs, sing an aria, or create a farm full of balloon animals, you can trust your lungs to supply all the breath you need.

Like your heart, your lungs drive one of the ceaseless rhythms of your body. Their quiet pattern of inhalation and exhalation starts with your first, gasping breath as a newborn, and continues unbroken until your last, ragged breath on your deathbed. In between is everything that matters.

In this chapter, you'll discover how your lungs collect the oxygen you need to turn food into fuel. You'll study the art of good breathing, and learn how to project your voice and stop snoring. Then you'll explore your upper airways, consider the benefits of nose breathing, and learn to love your nasal mucus. Finally, you'll meet a few of the dangers that threaten your respiratory system, from tobacco smoke to asthma.

The Respiratory System

Every minute, you inhale 10 to 20 times. The process is automatic, unconscious, and impossible to override for more than a few minutes.

The reason you spend so much time sucking up air is obvious—you need the *oxygen* it contains to live. But what, exactly, makes oxygen so vital to your body's health? To find the answer, you need to revisit your body's energy-production system.

As you learned in previous chapters, your body's principal source of energy is food—namely, fat and sugar, which your body breaks down to power its biochemical reactions. (Actually, you might say "sugar and sugar," because your body breaks down fat to release *glycerol*, which your liver then converts to *glucose*.) Every cell in your body uses sugar, passing it through a complex ladder of chemical reactions until it yields a tiny bit of energy.

But there's a catch: The complex biological processes that convert raw sugar into energy need lungfuls of oxygen to do their work. Without this ingredient, the whole series of reactions grinds to a halt. Your cells can get by for a brief time by generating energy through the less efficient process of anaerobic reactions (see page 73), but soon after, they stop working and begin to die. Your brain, with its high energy requirements and limited energy stores, is the first to go.

> **Note** Your body's insatiable need for oxygen is one shortcoming that evolution never corrected. A better design would keep your miraculously efficient lungs but add a back-up oxygen-storage system. That way, leaving a child in a bathtub or inhaling a grape wouldn't be life-threatening.

The organs that handle this critical task of oxygen collection are your lungs. In the next section, you'll take a closer look at how they work.

The Structure of Your Lungs

From the outside, your lungs look and feel like a pair of soft, pink sponges. Together, they fill most of your chest. While you might think they're mirror images of each other, the lung on your right is slightly larger than the one on your left. That's because your left lung needs to leave room for your heart.

On the inside, your lungs are far more complex. The passageway that leads down into them, called the *trachea*, divides into a complex, inverted–tree-like structure. These airways are lined with smooth muscle and reinforced with rings of cartilage. Each branch becomes progressively smaller until it terminates in a tiny tube called a *bronchiole*, which is barely the width of a single hair.

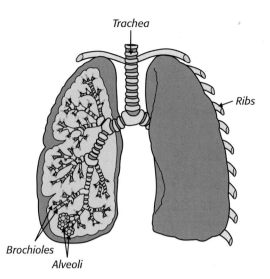

At the end of your bronchioles are tiny air sacs called *alveoli*, which are arranged in clusters, like grapes. A network of tiny blood vessels wraps each alveolus. These vessels are so narrow that your blood trickles through them just one cell at a time.

It's easy to underestimate just how many thin, branching passageways fill your lungs. All together, your lungs hold a staggering 1,500 miles of tubing. Solid estimates suggest that the total surface area of your alveoli is the same as that of a standard tennis court. That's important, because your lungs' surface area determines how quickly you can absorb oxygen.

> **Note** The experts agree—your lungs are far more powerful than they need to be. Unless you're an elite athlete, even intense exercise won't tax your lungs. You're far more likely to run into the limits of how fast your body can distribute and process oxygen than how fast it can absorb it. That's why smokers can nurse their pack-a-day habit for years without having to struggle for breath. Even as they fill their lungs with tar, they still have enough capacity to absorb all the oxygen their bodies need—for the time being.

The Great Gas Exchange

Your body's all-important gas exchange takes place in the alveoli—those grape-like sacs at the end of every passageway in your lungs. Thanks to these microscopic pouches, your body extracts all the oxygen it needs out of each breath you take. It's a remarkable feat, considering that the air stays in your lungs for just a few seconds.

Actually, two essential operations take place in your lungs: *Oxygen* passes through the thin walls of your alveoli and dissolves into your blood, and *carbon dioxide* makes the reverse trip, passing out of your blood and into the air in your lungs so you can exhale it. (Carbon dioxide is a byproduct of your body's energy-production process. It's toxic, and removing it from your body is nearly as important as absorbing oxygen.)

Note You might assume that your lungs strip all the oxygen out of the air you inhale. In truth, they remove only a small quantity. That's why mouth-to-mouth resuscitation works—even when you breathe your oxygen-diminished air into someone else's lungs, there's still plenty of oxygen left over for that person's body to claim.

This process—swapping oxygen and carbon dioxide—relies on a critical protein called *hemoglobin*, which acts as a super-efficient transport vehicle. Each *red blood cell* holds two or three hundred hemoglobin molecules, and each of these can latch onto four oxygen molecules as it runs through your lungs.

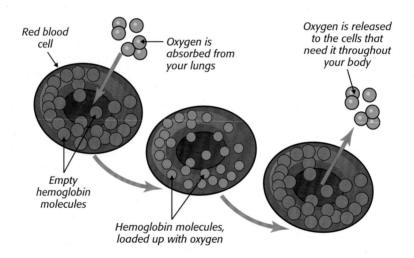

Red blood cell

Oxygen is absorbed from your lungs

Oxygen is released to the cells that need it throughout your body

Empty hemoglobin molecules

Hemoglobin molecules, loaded up with oxygen

In many respects, your lungs work like a giant loading dock. As blood circles past the millions of alveoli in your lungs, hemoglobin scoops up the oxygen in the air you just inhaled. The hemoglobin then continues on its way, eventually distributing this newly acquired oxygen to every cell in your body. Hemoglobin works a similar, but complementary, magic with carbon dioxide. It mops up stray carbon-dioxide molecules throughout your body and transports them to your lungs, where it releases them.

The Practical Side of Body Science

Holding Your Breath

In the short term—say, two or three minutes—you have complete control over your ability to hold your breath. But try to hold it for much longer and you'll run into your body's stubborn survival instinct.

As you hold your breath, the level of oxygen in your blood begins to sink. At the same time, the level of carbon dioxide—the waste left over from your body's energy-generating operations—starts to rise. Interestingly, it's the change in the carbon-dioxide level that your body really notices. Your brain detects this increase and eventually triggers an overwhelming urge to resume breathing.

Reckless divers sometimes take advantage of biology to increase their breath-holding time by hyperventilating before they dive. (To *hyperventilate* is to suck in more air than you actually need, usually by taking quick, deep breaths.) After hyperventilating, the diver feels a reduced urge to breathe and can stay underwater longer. This isn't because he's loaded up on oxygen (he isn't). Rather, it's because he's lowered the amount of carbon dioxide in his body. As a result, it takes longer to trigger the brain's safety mechanism that forces the diver to start breathing again. This practice is dangerous on several counts. The most obvious risk is that both oxygen deprivation and carbon dioxide overload lead to dizziness and can cause a blackout, which will leave the diver lying quietly unconscious at the bottom of a body of water.

Incidentally, the current record for breath-holding is a staggering *17 minutes*, set by the endurance artist David Blaine. To pass the 5-minute mark, Blaine employs several specialized techniques. Most important is the *mammalian diving reflex*—an evolutionary throwback that causes the human body to lower its heart rate and blood pressure quickly when submerged in cold water, thereby conserving oxygen. Also important is inhaling pure oxygen rather than air (which allows practiced breath-holders to double their breath-free time) and using a trick called lung packing (swallowing hard to force more air into the lungs). All the rest is practice and masochism.

Breathing

Your lungs don't breathe by themselves. Instead, a sheet of muscle sitting just under your lungs powers every breath you take. This muscle is called the *diaphragm*.

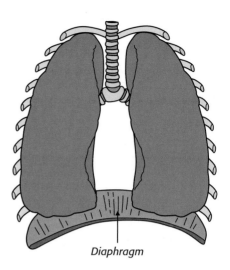
Diaphragm

Whether you realize it or not, you inhale by contracting your diaphragm. This contraction pulls the dome-shaped diaphragm muscle down, expanding your lungs so that air rushes in. Although the air travels through your mouth or nose, these body parts don't play an active role in breathing. In other words, whether you sniff with your nose or suck with your mouth, your diaphragm does the real work of inhalation. (The chest muscles around your ribs help out a bit, too.) To feel your diaphragm in action, place your hand over your stomach while you inhale.

Breathing Exercises

You've probably noticed that you don't need to micromanage your diaphragm. Whether you're awake, asleep, or sitting in a Monday-morning meeting in some indeterminate state, your diaphragm keeps up its tireless rhythm. However, many experts recommend *breathing exercises* as a way to get in touch with what your diaphragm is doing. They suggest that conscious breathing can make you feel more energetic or help you find inner peace.

To some, the whole idea sounds a bit like a New Age nightmare. But you don't have to don a magic magnetic bracelet or carry around power crystals to enjoy the benefits of a good, deep breath. In fact, breathing exercises can help reduce stress, treat anxiety disorders, manage the symptoms of certain lung diseases, and help women bear the pains of labor. Furthermore, controlled breathing is an essential foundation for other practices that use the lungs, like singing and public speaking. And it's a cornerstone of the exercise-and-meditation fusion called *yoga*.

The best breathing exercise to try first is *deep breathing*, a practice described in the next section.

Taking a Deep Breath

If you're like most people, when you breathe in, you take in a tiny puff of air that fills only the top of your lungs. You usually keep your diaphragm clamped firmly in place with tense, tight abdominal muscles. You hold your breath while concentrating. And when you try to take a deep breath, you puff out your chest without making the most of your diaphragm or filling your lungs.

The first step to deep breathing is awareness—in other words, knowing that your diaphragm exists. Follow these steps to take a truly deep breath:

1. Start by sitting comfortably or lying on the ground (which is even better). Make sure you aren't wearing restrictive clothes, like tight-fitting jeans, that will hamper your breathing.

2. Put one hand on your chest and the other on your stomach. This helps you *feel* where you're breathing, which is a critical part of getting it right.

3. Inhale through your nose, *slowly*. This process should take about five seconds. As you inhale, you should feel your stomach expand and your hand gently rise. As you continue inhaling, you'll feel your chest expand.

4. Exhale through pursed lips, *slowly*. (The pursing is to discourage your natural instinct to blow out the air rapidly.) The exhalation process should take longer than the inhalation step (say, 7 or 8 seconds). It may help to count the seconds in your head.

5. Rest and repeat 5 or 10 times.

You may want to run through this sequence when you first wake up in the morning (which gives you a great excuse to stay in bed for a few more minutes) and before you fall asleep at night. When you get comfortable with it, try using it to ground yourself during times of stress. You can also use this breathing exercise as a stepping stone to other pursuits, like transcendental meditation or vocal training.

The Path to Your Lungs

Although the real action of gas exchange takes place in your lungs, they're only part of your respiratory system. Before air can make its way into the thicket of branching passages that fill your lungs, it first needs to pass through your upper-respiratory system (which is shown on the next page).

Common Conditions

Snoring: The Nightly Thunder

A great many pleasant sounds can come out of a person's airways, but snoring isn't one of them. In fact, loud snoring can generate the same volume as an average lawnmower (which you definitely wouldn't invite into bed with you). For this reason, the effects of snoring are usually felt first by whoever has the misfortunate of sharing your bed.

Technically, snoring occurs when the passageways at the back of your throat become partially blocked, causing the soft tissues in and around your throat to vibrate noisily. Nearly half of all adults have an occasional snoring episode, and about a quarter are chronic snorers. On its own, snoring may be harmless (for the snorer). However, if you bed down with a snorer, you may suffer from interrupted sleep, stress, and even hearing loss.

Furthermore, snoring can be a sign of a much more serious problem, called *sleep apnea*, where the sufferer repeatedly stops breathing during sleep. With serious sleep apnea, interruptions can occur more than a dozen times per hour and can last 1 or 2 minutes. The result is that the sufferer never gets proper sleep, spends most of the day in a dangerous haze, and is at increased risk of heart disease.

If you're an occasional or frequent snorer, avoid the three most common aggravating factors: smoking, being overweight, and quaffing alcohol in the hours before bed. (Alcohol relaxes the muscles in the back of your throat, making it easier for them to flap loudly together.) You can also try sleeping on your side—a simple switch that immediately reduces the nightly emanations of many a problem snorer. (To train yourself to side-sleep, take a shirt that has a pocket in the front, put a tennis ball in the pocket, and wear the shirt backwards.) If you're still suffering, or you suspect sleep apnea, contact a doctor who can investigate. Possible remedies range from surgery to a dental appliance that you wear while you sleep.

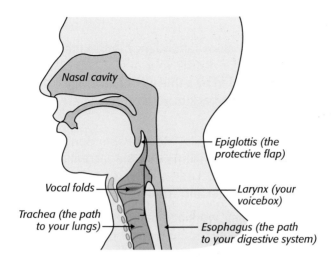

Nasal cavity

Epiglottis (the protective flap)

Vocal folds

Larynx (your voicebox)

Trachea (the path to your lungs)

Esophagus (the path to your digestive system)

Air enters your respiratory system through one of two passages: your nose or your mouth. Ideally, you breathe through your nose (or at least inhale through it, which is the most important part). That way, you can let your mouth take care of more glamorous tasks, like talking, eating, and amorous lip locking.

There are several reasons to favor your nose:

- **Air conditioning.** As you draw air through your nasal cavity, you warm and humidify it. This conditions the air so that it matches the environment of your warm, moist lungs.

- **Sense of smell.** As you learned on page 124, your nasal cavity holds a magic strip of smell receptors. They let you monitor the air for odors as you breathe.

- **Nighttime indignities.** Breathing through your mouth at night increases the likelihood you'll snore. It also dries up your saliva, allowing mouth bacteria to flourish and giving you bad morning breath.

- **Filtration.** Most important, dirt-trapping hair fills your nostrils and mucus lines your nasal cavity. As air travels through your nose, these natural filters trap some of the particles that might otherwise harm your lungs (page 150). They can also slow down the bigger particles that *do* travel into your lungs, so they don't penetrate as deeply.

Of course, if allergies or a cold seal your nose shut, you'll need to make do as a temporary mouth-breather. You may also need to breathe through your mouth if you have a structural nose problem (like a deviated septum) that prevents you from breathing comfortably through your nose. However, this disorder can often be corrected through surgery.

The Nasal Cycle

Your nose has a bit of built-in redundancy. Although you could survive perfectly well with a single nostril, your nose holds two parallel breathing passages that are divided by a thin wall of cartilage called the *septum*. These passages meet at the back of your throat, where they take a single tunnel down to your lungs.

So why do you have two nostrils? Most people don't spend much time comparing them, and simply assume that the second one is there to take over when the first one is blocked—say, by a nasty cold or an antisocial finger. But the reality is subtler and stranger.

Your two nostrils shift their workload back and forth in a delicate dance called the *nasal cycle*. At any moment, most of the air you inhale travels through just one nostril, while a much smaller amount seeps in through the other. At some point, the nasal cycle reverses course and the workload shifts to the other nostril. The length of time between nostril switching varies, depending on the individual and various other factors, but each cycle usually lasts from 40 minutes to several hours. (That's why you often have intermittent periods of easy breathing even when you suffer from a heavy cold. In this case, one nostril is plugged more than the other.)

> **Note** It's easy enough to find out which nostril is currently in control. Cover one nostril and breathe in and out, then do the same with the other nostril. When it takes more effort to breathe, you're covering the dominant nostril.

For the longest time, scientists had no idea why the nasal cycle took place. After all, your nostrils aren't doing much work, so there's no reason they need hours of idle time. But recent research has uncovered the apparent reason: nostril switching improves your sense of smell.

To understand why, you need to realize that the nasal cycle changes the way air passes through your nose. In your dominant nostril, the air moves very quickly. In your other nostril, it seeps through more slowly.

This difference is important because odor-causing chemicals vary in the amount of time they take to dissolve through the mucus that lines your nasal cavity. Chemicals that dissolve quickly have the strongest effect in a fast-moving airstream that spreads them out over as many odor receptors (page 124) as possible. But chemicals that dissolve slowly are easier to smell in a slow-moving airstream. If the air rushes by too quickly, the chemicals will be whisked away before they've reached any odor receptors. That's why your nose has both a fast road and a slow lane. Quite simply, the combination of two nostrils with different airflows gives you a more detailed "smell picture" of the world around you.

Incidentally, your nose performs nostril switching using special erectile tissue that swells to narrow the passageway in the nostril that's taking a break. Collectors of bizarre body trivia take note: It's the same sort of erectile tissue that's at work in male and female reproductive organs.

The Dangerous Crossing

One of the human body's most notorious limitations is the way your feeding chute (your *esophagus*) passes just over your breathing tube (the *trachea*). A flap of cartilage called the *epiglottis* closes over your airway when you swallow to prevent food from slipping in (see the figure on page 141).

The problem is that the epiglottis isn't foolproof. If you cough or try to talk while eating, you can force your epiglottis open. With bad timing and the wrong food, you can end up blocking your airway completely.

Life-threatening choking is distinctive: Silence is the most obvious symptom. Because people use the same passageways to breathe and talk, someone with a complete airway obstruction will be almost completely unable to speak, gurgle, cough, or wheeze.

You know what comes next—the *Heimlich maneuver* (also known as *abdominal thrusts*), as you've seen in countless television movies. But don't act yet, because modern first-aid procedures suggest this sequence:

- Encourage the person to try to cough.

- Slap the person's back 5 to 20 times, using the heel of your hand. The goal is to create vibrations that can dislodge an obstruction.

- Now it's time for abdominal thrusts—a risky but potentially life-saving move. Stand behind the person and wrap your hands around his abdomen, just above the belly button. This is where you'll find the diaphragm (page 138), the muscle that sits just under the lungs and controls breathing. Make a fist with one hand, and wrap your other hand around it. Then pull your fist upward and inward, forcefully. You may need to repeat the move several times. Done right, abdominal thrusts increase the pressure in the lungs and hopefully blow out the lodged object. (They may also bruise the abdomen and fracture a rib.)

Your Mucus

One of the most notorious parts of your upper-respiratory tract is *mucus*, the thick, slippery coating that lines your breathing passages all the way to your lungs.

Mucus scores high on the list of Topics to Raise If You Never Want to Be Invited Out Again. However, biology nerds know there are many reasons to love your mucus. It plays essential defense and clean-up roles in your respiratory system, trapping nasty substances—from pollen and dust to bacteria and viruses. Without mucus, your nose would be in a perpetual state of dry discomfort, and you'd certainly be at more risk for nose and throat infections.

Many people associate mucus with colds and respiratory problems. While these conditions may cause your body to produce more mucus, they're more likely to thicken your existing mucus and inflame your airways, disrupting your body's normal mucus-disposal system and making you increasingly aware of the slimy stuff.

Tip When faced with unpleasantly sticky mucus, you may get some relief by increasing your fluid intake.

If you're an average person, your respiratory system produces a healthy four cups of mucus over the course of a day. Ordinarily, the mucus in your nose collects airborne contaminants and flows silently down the back of your throat and into your stomach. Your stomach then digests this mucus-coated garbage. This design highlights the power of your stomach's digestive juices. It also suggests that, when faced with dirt and disease-causing compounds, it's often safer to eat them than to inhale them.

Mucus Myths

Mucus has a commanding place in our collective imaginations. Although you may have heard some of the following factoids about the sticky stuff that lines your airways, there's not much more than a good story to back them up.

- **Milk thickens mucus.** People believe this bit of biological trivia so strongly that they really *do* report thicker mucus after drinking dairy products. But close analysis by mucus-ologists has found no change in the amount of mucus or its consistency after drinking milk. This myth may be partly reinforced by the creamy mouth-feel of milk, which can leave a coating on the tongue and throat.

- **Green mucus signals a bacterial infection.** Ordinarily, mucus is thin and clear. As your body battles an infection of any kind—bacterial *or* viral—your mucus will thicken and become yellow or green. (The green color is actually from an iron-containing enzyme your immune system uses for its germ-fighting reactions.)

- **Blowing your nose to expel mucus shortens a cold's duration.** Keeping a clean nose may reduce the chance of transmitting your cold. (Or not, as your immune system may have already neutralized the virus.) However, forceful nose blowing can drive viruses and inflammatory substances into your otherwise germ-free sinuses, possibly setting the stage for a sinus infection.

Your Voice

The human body is obsessed with reuse. You may think you're pretty clever when you manage to surf the Internet on your TV, but that's just peanuts compared to the multitasking prowess of your body. You already know that your senses often do double duty—your nose helps you smell and taste, and your ears let you both hear and balance. Similar reuse is at work in the circulatory system, your bones, and just about everything else in your body that's more complex than an eraser head. The supreme joke just might be the way the male body reuses its waste disposal pipe as a baby-making instrument.

> **Note** Incidentally, there's an important principle of biology that underlies this sort of reuse. It's called *exaptation*. Essentially, exaptation means that it's easier for evolution to repurpose an existing body part than to create something genuinely new.

One of the best examples of biological reuse gone wild is found in your upper airways. Your body uses the same entryway and parts of the same passage for no fewer than three distinct tasks: breathing, eating, and talking. (A combination of the first two can be fatal. A mix of the latter two can be just as harmful to your romantic life.)

The pairing of breathing and talking is particularly clever, because both require a steady flow of air. It's sort of like playing a bagpipe through your home's heating vents. As you breathe out, air flows from your lungs, and you can use it to speak and sing.

The Vocal Folds

The biological machinery that lets you speak is as amazing as any other superstar body part. Your voice starts to take form in your *larynx*, which is part of your neck—to find it, place your hand over the front of your neck and feel for a bulge. In men, this bulge is called the Adam's apple.

Your larynx houses two pliable pieces of tissue called *vocal folds*. (In the past, before scientists fully understood their operation, vocal folds were called vocal cords.) The vocal folds stretch over the top of your trachea— the tube through which your every breath must pass. When you breathe, the vocal folds separate to let air pass through. When you swallow, the vocal folds close tightly to prevent food from slipping down into your airways or lungs. This design hints at the evolutionary ancestry of vocal folds— they developed first as a safety mechanism to prevent choking, and only later became an instrument for producing sound.

Looking down your throat to the vocal folds

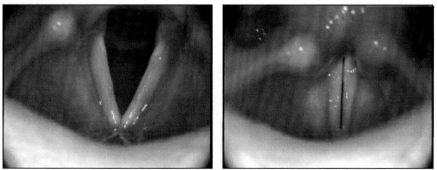

Used with permission of Dr. Christopher Chang with Fauquier ENT Consultants (www.FauquierENT.net)

*Open vocal folds
for breathing*

*Closed vocal folds
while swallowing or voicing*

To speak, you shut your vocal folds (but not as tightly as when you swallow). You then blow the air out of your lungs and against the folds, which makes them vibrate hundreds of times each second. At the same time, muscles in your larynx pull the folds, changing their tension and shape to vary the pitch of the sound you produce. Lastly, your tongue and lips fine-tune the sound with careful articulation. To utter a complete sentence, your brain must orchestrate a combination of quick, precise, and synchronized muscle movements. From the standpoint of the average speaker, the whole process is amazingly automatic.

> **Note** To watch some amazing videos that show vibrating vocal folds in action, send your favorite Web browser to *http://tinyurl.com/ceb38y*.

To make louder sounds when you speak or sing, you push more air along your trachea and through your vocal folds. If you do this for a long time, you can temporarily inflame the folds, making it difficult to speak—a malady called *laryngitis*. Frequent vocal-fold abuse can cause tiny growths called *vocal nodules* to form where the air pressure is highest. The result is a voice that's permanently breathy, low, or hoarse.

> **Tip** If your livelihood depends on speaking or singing, consider studying with a skilled vocal coach, who can catch bad habits and teach you practice exercises. For example, public speakers can learn vocal techniques to command attention and project their voices without shouting. Singers can learn to amplify and intensify their voices using resonance in their nose and mouth cavities (which is sometimes called the "head voice"). Either way, you'll be able to rely on your voice without risking vocal-fold damage.

Improving Your Voice

It takes little more than a 10-dollar tape recorder to see that most people aren't comfortable with their own voices. All your life, you listen to your voice from the inside, where it benefits from the booming resonance of your head. So it's little surprise that when you record your voice from the outside and play it back, the result is a somewhat tinnier sound.

The resulting insecurity makes some people reluctant to speak up and project their voices in crowded places. It encourages others to retreat into the comfort of mumbles. Fortunately, there's a solution. A bit of vocal practice can help you unlearn these bad habits and reclaim your voice.

In the following sections, you'll walk through a handful of fun, practical ways to improve your speaking, whether you're chairing a company meeting or cruising a nightclub for a hot hookup.

Get comfortable

If you aren't comfortable with your voice, no one else will be, either. So start by recording your voice in ordinary conversation and taking a close listen. This process has two key benefits: First, it lets you objectively hear the strengths and weaknesses of your speaking tone, so you can learn to improve it. Second, it helps you get accustomed to hearing your own voice. Over time, this can make you comfortable enough to speak up.

You can use the recording trick to help you work alone on the following exercises, or you can team up with a partner and do them together.

> **Tip** Many people automatically lower their voice in public places out of a misguided sense of decorum or privacy. But a confident speaker maintains a comfortable, consistent tone and doesn't worry about being overheard.

The man in the mirror

In your quest to refine your voice, it often helps to work on something that seems unrelated but isn't—your posture. Speakers who slouch tend to talk downward, take shallow breaths, and lose confidence—all characteristics that rob your voice of its projecting power.

To keep an eye on yourself in real time, you can try an old radio trick—speaking in front of a mirror. Then, to boost your speaking energy, practice talking with an absurdly exaggerated grin on your face.

Slow speech

To practice this exercise, read the same paragraph from a book or newspaper several times. Each time, try to speak a little more slowly, while enunciating as clearly as possible (exaggerate if necessary). Continue doing this until you find the absolute slowest speed at which you can speak without sounding like you're dozing off.

Slow speech is a cornerstone of good oratory. Barack Obama has a notoriously relaxed words-per-minute rate, for example. Slow speech helps prevent you from constricting your voice, raising its pitch, or mumbling. A related skill is *pausing*, which experienced speakers use to shape long speeches into careful paragraphs that allow listeners to follow along easily.

Many people who are uncomfortable speaking are even less comfortable pausing. It's as though they're afraid to give listeners the chance to reflect on their spotty oratorical skills. If this describes you, practice deliberately holding your pauses for longer than seems natural—say, a five-count.

Serial emphasis

Pick a simple sentence like "I never said he loved the maid." Repeat the sentence, emphasizing the first word, then the second word, then the third word, and so on. For example, "*I* never said he loved the maid. I *never* said he loved the maid. I never *said* he loved the maid. I never said *he* loved the maid," and so on.

Exaggerate the difference to dramatize the sentence and change its meaning. This exercise helps boring speakers break out of their lazy monotone and wake up to the power of varied expression. A similar exercise is to take a well-known nursery rhyme and repeat it using different personalities (for example, using a tone that's angry, delighted, confused, anxious, and so on).

Project the distance

Here's a good group exercise to try in a large room: Organize people into pairs and arrange them in a circle or two lines so that the partners are facing each other. Ask the partners to have a conversation about something straightforward—for example, their day, current affairs, or a recent movie they saw. As they speak, call out "Back!" every 20 seconds, at which point everyone must take a step back. The challenge is for everyone to continue their conversation without being drowned out or distracted by the other conversations, and without resorting to shouting. (Making sure no one trips over a stray piece of furniture is also a good idea.)

At the very end, when the partners are as far away from each other as possible, go through the pairs one by one. Each pair must then rush back together, holding their far-apart volume, and continue to talk for an additional 20 seconds at close proximity. The difference between their starting volume and their new, projected voice will be dramatic.

Pollution

When you think about pollution, you probably imagine belching smoke-stacks and clouds of car exhaust. But most of the airborne particles that threaten your lungs are invisible. These silent drifters, called *particulate matter*, fill the air around you, whether you're walking outdoors or sitting in the comfort of your home.

In the following sections, you'll take a closer look at these microscopic particles. You'll learn how they attack your lungs, and you'll pick up a few practices to help control the risk. Finally, you'll consider the very worst substance that humans regularly inhale—cigarette smoke.

The Size of the Problem

Unsurprisingly, particulate matter consists of tiny particles—vanishingly small bits of material formed through chemical reactions (say, burning fossil fuels) or shed from larger objects (like dust, pollen, and mold). Traditionally, environmental scientists distinguish between two broad categories of particulate matter, based on size:

- **PM10** (pronounced "pee em 10"). These particles are smaller than 10 microns across, but larger than 2.5 microns. (For comparison, a human hair is about 60 microns thick and the finest beach sand is 90 microns.)

- **PM2.5.** These particles are the really tiny ones—they're smaller than 2.5 microns thick.

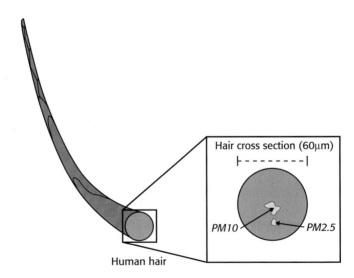

Hair cross section (60μm)

PM10

PM2.5

Human hair

Particles in these two categories act quite differently. PM2.5 particles are extremely light. They can travel hundreds of miles and stay aloft for weeks. PM10 particles can manage a few miles and a few hours, at best. But the most important distinction is the way these two particle types affect your lungs.

When you breathe, you suck in lungfuls of air and particulate matter. Your respiratory system's natural defenses trap many of these particles. They stick to the mucus-lined airways in your lungs. Once immobilized this way, tiny hairs called *cilia* slowly brush them back upstream and out of your lungs, so you can cough them out or swallow and destroy them with the digestive acids in your stomach.

But here's the catch: The tiny PM2.5 particles travel deeper into your lungs. They're less likely to get stuck in your passageways and more likely to travel all the way to the alveoli, where your blood absorbs oxygen. This poses a problem, because your alveoli aren't able to sweep out foreign particles like the rest of your lungs can. Instead, the particulate matter can remain lodged in your alveoli indefinitely, eventually damaging lung function and contributing to diseases like emphysema and lung cancer. Furthermore, some substances can pass through the walls of alveoli and directly into your blood. This is another case of seriously bad news, because PM2.5 particles are more likely to include toxic substances like lead.

> **Note** The air also holds plenty of particles larger than PM10—for example, visible dust and skin flakes. However, air-quality experts ignore these particles most of the time because you're much less likely to inhale them deep into your lungs, and your body's defense systems (like the hairs of your nose) are much more likely to catch them.

To take good care of your lungs, you should reduce your exposure to PM2.5 particles as much as possible. If you live in a large city, pay attention to the *air-quality index*, which soars when the air is thick with certain pollutants (including ozone, carbon dioxide, and particulate matter). When the index is high—often during hot, humid weather—the outside air is most likely to harm your lungs and aggravate medical conditions. On bad air days like these, just stay in. And whatever you do, avoid vigorous exercise when pollution is high. Many air experts believe a morning jog through the smog does more cardiovascular harm than good.

Indoor Air Quality

If you're like most people, when you worry about pollution, you think about the particles in the air outside. After all, that's where cars drive, dust blows, and factories do their dirty work. But air-quality experts know the grimy truth—when it comes to air pollution, your lungs may have more to fear indoors than they do outside.

If this seems contradictory, you need to take a closer look at the environment where you spend most of your time. Items that fill many modern homes, such as carpeting, furniture, paint, and cleaning supplies, can release *volatile organic compounds* (VOCs). In addition, biological sources such as dander from your pets, skin flakes from your body (page 10), and spores of mold contribute to indoor air pollution. And because modern homes are surprisingly airtight, particles can build up to concentrations you'd never face outdoors. Add to this the fact that the average person spends 90 percent of the day inside, and you can see why your lungs have more to worry about from a relaxing day at home than a walk around the block.

> **Note** One common indoor air pollutant is the residue left from cigarette smoke. Health researchers have coined the term *third-hand smoke* to describe the toxic particles that remain long after the visible smoke has drifted away, clinging to furniture, carpeting, upholstery, and clothing. New, but somewhat controversial, research suggests that exposure to this residue can be damaging. It's a particular concern for young children, who are likely to crawl along particle-laden carpets.

So what can you do to improve indoor air quality and reduce your exposure to PM2.5 particles? Here are some proven techniques:

- **Ventilate.** Opening a window is one of the easiest ways to clear out indoor air pollutants. Even without a strong breeze, built-up pollutants will naturally spread out and drift out of an open window. Of course, this technique isn't as useful on a hot and smoggy day, or if you live near a pollution source (say, a few feet from a heavily trafficked road). In cases like these, you might get more mileage out of an air exchanger, which brings in outside air, filters it, and uses it to heat or cool your house.

> **Tip** A regular dose of fresh air keeps your indoor environment healthy. Ventilation is particularly important when you do something that releases large amounts of indoor pollutants, such as painting (even with low-VOC paints) and using strong cleaning products (like those you use to clean the bathroom, floor, oven, and so on).

- **Use exhaust fans.** Stovetop cooking creates plenty of potential lung irritants, and our hot, steamy showers generate the humidity that allows mold to thrive. To cut down on these sources of indoor air pollution, make sure you have an exhaust fan in every kitchen and bathroom. Use the kitchen exhaust fan while cooking, and use the bathroom exhaust fan while cleaning and after bathing.

- **Air out.** Air-quality experts recommend that you air out problem items before you bring them into your house. This includes dry-cleaned clothes and new carpet, both of which release hefty quantities of VOCs. (New carpet will probably continue releasing VOCs 2 or 3 years after installation, but you can cut down on the intense initial exposure by giving it a couple of days to air out—or better yet, by going with hard floor coverings, like wood.)

The Practical Side of Body Science

Second-Rate Cleaning

Some of the practices that people take to improve indoor air quality don't have the effect you might expect. Here are some examples:

- **Air cleaners.** Air cleaners can trap indoor air pollutants, but many don't work that well, and some emit extra ozone that can aggravate asthma. If you plan to buy an air cleaner, check the reviews from a reliable source like *Consumer Reports* (*www.consumerreports.org*). And remember that air cleaners have limitations—for example, a good one can remove particles from nearby air, but it can't purify the whole house. (In fact, because of the way that extremely light PM2.5 particles drift through the air, an air cleaner might not be able to keep even a single room particle-free.)

- **Vacuuming and dusting.** These activities are keenly important to prevent excessive dust from building up in your home. However, both practices disturb dust and send particles into the air, which means that air quality may actually *decrease* immediately after a thorough cleaning. To reduce this effect, dust with a damp rag or an electrostatic cloth. You can also use a vacuum that has a built-in HEPA filter, which traps fine particles that would ordinarily be blown back into your house as part of the vacuum's exhaust. (However, most vacuums are so poorly sealed that plenty of dust can escape, no matter what type of filter they use.)

- **Air fresheners.** Air "freshening" is an odd idea, with roughly the same scientific underpinnings as palm reading. Many air fresheners simply use aromas to mask offensive smells, although some include a nose-numbing chemical that makes it more difficult for you to smell anything, good or bad. Along with these dubious ingredients, air fresheners release several chemicals that are linked to lung irritation and (in high concentrations) to lung damage.

The Riddle of Asthma

In the modern world of cigarettes, automobiles, and heavy industry, it seems hardly surprising that people face increasing rates of lung diseases. However, it's surprisingly difficult to nail down the cause of *asthma*—a chronic condition that causes sudden, unexpected narrowing in the airways of the lungs.

Contrary to what you might think, asthma isn't caused by toxins in the air, but from your body's overactive response—often, to otherwise minor irritants like dust, mold, and animal dander. During a severe asthma flare-up, the muscle tissue that lines the airways in your lungs can become so swollen that it chokes off your air supply.

The mystery of asthma is why it's so common in the rich, industrialized countries of the Western world. The reason remains elusive. Some suggest that it's a side effect of modern living in tightly closed spaces (and, by extension, the build-up of indoor air pollutants). Other researchers support the *hygiene hypothesis*, which argues that our clean living habits limit our exposure to infection and allergens early in life. (You'll learn more about this line of thinking on page 232, when you tackle your body's immune system.) Current research is contradictory—some studies suggest that early exposure to allergens can trigger lifelong asthma, while others point to lower asthma rates in environments filled with germs and irritants (daycare centers, families with many children, crowded living spaces, and so on).

Whatever the case, if you experience symptoms of asthma, such as occasional trouble breathing or a nagging shortness of breath, head to your doctor for a lung-function test. Even if your asthma isn't severe enough to cut off your oxygen supply, the prolonged inflammation can gradually damage your lungs if you don't manage your asthma with medication. (And no, asthma inhalers cannot cause obesity, even if they contain steroids. They act directly on the lungs to expand air passageways or reduce inflammation.)

Tip If you have asthma, you may be able to reduce your symptoms by controlling house dust and indoor air pollutants. The key is to identify your personal asthma triggers and work to reduce them.

The Contents of Cigarette Smoke

No chapter about lungs would be complete without mentioning the swiftest and most effective way to damage them—by inhaling generous quantities of toxic cigarette smoke. While anyone born in the era of color television already knows that cigarettes are poison in a stick, you may not know exactly how they damage your lungs.

It turns out that there is no single answer. Although it's tempting to talk about cigarette smoke as though it's a single thing, it's actually a lethal brew of more than 4,000 lung-scarring chemicals, with nearly 100 known *carcinogens* (cancer-causing agents) in the mix. The following figure highlights some of the nasty substances you'll encounter in cigarette smoke, and links them to the industrial products they're more often associated with.

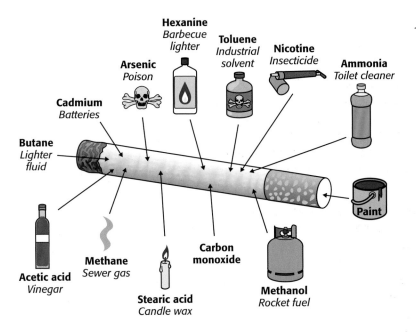

While cigarette smoke is hard on your lungs, its most popular ingredient—nicotine—is a disaster for the rest of your body. It's particularly hard on your heart, as it increases your heart rate while constricting your arteries, leading to higher blood pressure and hardening of the arteries. That's why smoking is so tightly associated with heart disease (see the box on page 172). The brutal truth is this: Every cigarette you smoke inches you closer to a coronary disaster.

Because human lungs are somewhat overbuilt with capacity to spare, smokers don't suffer the most damaging effects of cigarette smoke for some time. However, it doesn't take many packs of puffing to coat the lungs with a rich assortment of tars. If you're a regular smoker, you'll have no trouble picking out your lungs in the picture below, which is featured on cigarette warning labels in several countries.

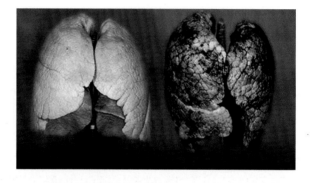

The Practical Side of Body Science

How to Heal a Smoker's Lungs

Your lungs have a nearly miraculous ability to heal themselves. Although longtime smokers will always have an increased risk of lung cancer, kicking the habit brings immediate health benefits. Here are some benefits the average smoker can expect from quitting:

- After 8 smoke-free hours, your blood's carbon-monoxide and oxygen levels return to normal. This change can reduce fatigue and headaches, and increase your ability to exercise.
- After 1 smoke-free day, your chance of having a heart attack has already declined.
- After 2 or 3 smoke-free weeks, your lung function improves by nearly a third.
- After several smoke-free months, new and healthy cilia (the tiny lung-cleaning hairs) begin to grow, increasing your ability to clean your airways and prevent infection.
- After 1 smoke-free year, your risk of heart disease is half that of a continuing smoker.

7 Your Heart

I f your body were a movie set and all your organs were actors, your heart would be the overexposed star with the hotshot agent.

After all, your body is *full* of critically important pieces. Damage your diaphragm and you can't suck in a single breath. Remove your pancreas and you won't digest a solitary cracker. Lose your liver and you'll die within hours. Snip your spinal cord—well, you get the idea. Despite the obvious importance of all these body parts, they never get the media attention they deserve. Instead, they tick on quietly in the background, performing essential but unglamorous tasks, while the heart hogs the spotlight.

Given its star status, it's surprising that a lot of what the average person knows about the heart is, well, a bit off. When most people think of the heart, they see clogged arteries and chest-clutching heart attacks. But the biological truth is a little subtler and a whole lot messier. First, your heart doesn't run the show—it's actually just one piece in a complex, cooperating system. Second, your heart isn't always open about its problems. In fact, it might not warn you even when it's badly faltering.

In this chapter, you'll get up-close and intimate with your heart. You'll see how it works and what makes it fail. You'll walk through a scenario of a real-life heart attack, and learn to avoid the crucial mistakes that could cost you your life. Finally, you'll review the touchstones of heart health and learn how to track your heart rate so you can get the biggest possible boost from cardio exercise.

The Pump Inside

Buried in your chest, just slightly to the left, is the fist-sized pump that powers your circulatory system.

Before going any further, it's important to make one thing perfectly clear: Whatever characteristics your heart has, it most certainly is *not* heart-shaped. In fact, the origin of the pleasantly curved heart icon is a bit of a mystery. Popular theories link it to botched anatomy drawings, religious iconography, and the leaf of an ancient herbal contraceptive. (We'll let you connect the dots between true love and truly effective birth control.)

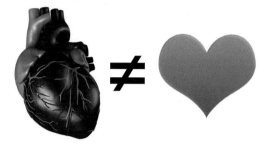

Your heart has both impressive abilities and undeniable limitations. Chief among its strengths is a tireless endurance that's unmatched by any other muscle. In fact, over the course of your life, your heart will deliver a couple of *billion* beats, all without pausing for a break. But your heart's most impressive quality is its flexibility—the way it can ramp up from slow, steady pulses to hard-driving contractions in mere seconds. Whether you're dozing on the couch or sprinting through an airport, your heart works at precisely the right rate, delivering a steady supply of blood and oxygen to keep your body working. Artificial hearts can't come close.

Your heart's chief weakness is its fragility. Because it's unable to replace dead tissue, any damage it suffers impairs its performance for life. From an evolutionary point of view, this drawback isn't a surprise. Without modern medicine, our ancestors were unlikely to survive any sort of heart attack, so they had little use for a heart that could rebuild itself after the event.

Your Heartbeat

Although you may think of your heart as a fleshy mechanical pump, it's actually *two* pumps in one. The right side of your heart pumps blood toward your lungs, where it loads up on oxygen. The left side of your heart pumps this oxygen-rich blood to the rest of your body to keep it alive. Although these two pumps beat in time, a thick wall of muscle separates them, and the blood in one half never mingles with the blood in the other half.

To better understand how your heart beats, check out these diagrams. They show the three stages of a single heartbeat:

1. Blood enters the first two chambers of your heart. On one side, blood goes into the *right atrium*. On the other side, it flows into the *left atrium*. These two chambers are completely separate, and your blood enters them through two different veins.

2. Your heart contracts its atriums, forcing blood into the two chambers just underneath. The blood from the left atrium enters the *left ventricle*, and the blood from the right atrium enters the *right ventricle*. Your heart controls the direction of flow using two tiny valves that separate the atriums from the ventricles—once the ventricles are full of blood, these valves snap shut.

3. Your heart then contracts its ventricles, pushing blood out of the heart and off on its journey, through two separate passageways. Once again, a pair of valves closes behind the exiting blood to prevent it from washing back.

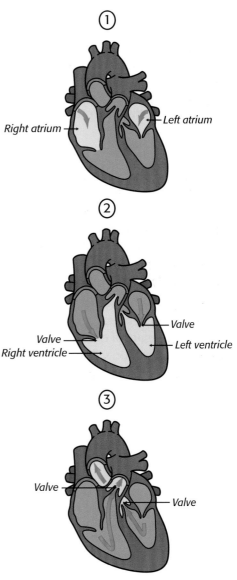

Most people don't realize that the heart's distinctive sound (the "lub dub" you hear when you listen to your heartbeat through a stethoscope) is made by your heart's valves. The first part of your heartbeat—the lub— occurs when blood passes from the two atriums into the two ventricles, at which point both valves close simultaneously. The second part—the dub—occurs when blood exits your heart and the second set of valves, the ones in your ventricles, snaps shut.

Doctors use a stethoscope to monitor these valves. If your heart has a structural abnormality or its valves aren't working properly, your blood will flow unpredictably, resulting in a variety of additional sounds, from whispers to clicks. Doctors call this group of sounds *heart murmurs*. Although some murmurs can reveal potential problems, most are harmless quirks.

Heart Rate

When you relax, your heart beats about 60 to 80 times a minute. (On average, a man's heart beats 70 times per minute and a woman's beats 75 times per minute.) If you're an athlete or a fitness geek, the training you do improves your heart's efficiency and helps it pump more blood with each beat. As a result, your resting heart rate may drop as low as 40 to 60 beats per minute.

If you're older and out of shape, your heart may be gradually weakening. To keep up with the demands of your body, it may beat faster, increasing your resting heart rate to closer to 100 beats per minute. A heart rate like this is a hallmark of poor fitness and a flashing danger sign of heart trouble ahead. Studies show that a high resting heart rate goes hand-in-hand with a higher risk of heart attack, especially later in life.

That said, there's a healthy amount of natural variation in each person's heart rate. So even if yours clocks in at 60 beats per minute and your friend's sets a quicker pace of 75 beats per minute, you can't assume that your heart is the superior specimen.

Note Unless your heart rate seems unnaturally fast or unnaturally slow, you're unlikely to give it much thought. Tracking your heart rate during exercise, however, is worthwhile, because it tells you how hard your heart is working. You can use this information to make sure you get the most out of your cardio workouts, as described on page 174.

Mind Control for the Heart

For the most part, your heart rate ratchets up and slows down on its own, without any conscious control on your part. However, an interesting technique called *biofeedback* can give you the ability to influence automatic body processes like heart rate.

In a typical biofeedback session, a trainer hooks you up to specialized equipment that measures something you can't ordinarily perceive, like brain activity, skin temperature, or muscle tension. The equipment then translates this information into something you *can* perceive, such as a tone that varies in pitch, a light that varies in brightness, or an image with shifting lines. You then follow a series of mental exercises until you eventually stumble on a technique that causes the right physical change and the corresponding auditory or visual signal. For example, a biofeedback session might measure the activity of a brain region known to influence heart rate. Through trial and error, you can learn how to slightly slow your rate using nothing more than mental power.

The ultimate goal of most biofeedback training is to give you the ability to consciously reduce your reaction to the stresses of ordinary life—for example, damping down a racing heart or a sudden spike in blood pressure—even without the help of the fancy biofeedback equipment. Despite its promise, however, biofeedback hasn't yet graduated from an experimental curiosity to a truly therapeutic tool.

Blood Pressure

To most people, heart rate doesn't seem like a terribly important number. That's OK, because a far more important measure of your heart's health is blood pressure—in fact, it's the most reliable predictor of heart trouble to come.

Blood pressure is the force your blood exerts on the walls of your arteries. Many factors influence blood pressure, including your heart rate, the amount of blood in your body, and how thick that blood is. The most significant thing your blood pressure measures is the *condition* of your arteries. As they age, your arteries stiffen and swell, forcing your heart to pump harder to deliver blood to the far corners of your body. Most people find that their blood pressure creeps up as they grow older. It's a matter of debate whether this is a natural consequence of an aging body, or the cumulative toll of a modern lifestyle.

Blood pressure can vary in different parts of your body, so doctors generally measure it in the main artery of your upper arm. A blood-pressure reading actually consists of two measurements. The *systolic* pressure is the force your heart exerts when it pushes blood out. The *diastolic* pressure is the force between heartbeats.

In the past, the medical community thought that diastolic pressure was the more important of the two, possibly because your blood vessels experience this level of pressure between heartbeats, which is most of the time. But recent research shows that systolic pressure is the best predictor of future ills. (It's also possible that the *difference* between the two figures is even more significant, which means that a person with a high systolic pressure and a low diastolic pressure is at the greatest health risk of all.)

The following table outlines commonly observed blood-pressure ranges and what they mean. Although the best way to measure your blood pressure is with the high-quality equipment in a doctor's office, you can perform an informal check with a free, automated machine at many drugstores.

Category	Systolic	Diastolic
Hypotension	Less than 90	Less than 60
Normal	90–119	60–79
Prehypertension	120–139	80–89
Stage 1 hypertension	140–159	90–99
Stage 2 hypertension	160 or greater	100 or greater

Slightly low blood pressure puts you at greater risk of fainting and dizziness, but it doesn't raise any health concerns. (However, *extremely* low blood pressure—a.k.a. hypotension—may be the sign of an underlying disease.)

High blood pressure isn't so forgiving. Called *hypertension*, it takes a slow and steady toll on your body, scarring your arteries and organs like a high-pressure washing hose that's set too high. Hypertension is insidious because it has no obvious symptoms until it's far advanced, and by then it may already have damaged your heart, eyes, kidneys, and even your brain.

Now that you know the importance of tracking your blood pressure, the next question is this: What can you do if it's outside the norm? If you already have hypertension, several lifestyle changes may help reduce your blood pressure (see the box on page 163). However, these steps are unlikely to reverse the condition completely. That means that once you make all the changes you can, you need to take careful blood-pressure readings and work with your doctor to choose the most suitable pressure-lowering drugs. And be warned: The cardiac unit of many a hospital is full of self-medicating patients who tossed their drugs after adopting a "miracle diet" or a serious workout regimen, only to suffer devastating heart problems. A healthy lifestyle is an admirable thing, but it's not likely to repair weakened arteries or roll back time.

If you're in the warning zone, with a systolic blood pressure that's nudged past 120 but still sits under the 140 mark, your prospects are better. By making the right changes, you can delay the onset of hypertension—perhaps indefinitely. Even if you eventually cross the dreaded threshold of hypertension, your hard work can keep you off blood-pressure medication for years before that point.

Battling Hypertension

As you learned earlier, you can tackle high blood pressure in two ways—through lifestyle changes and with a slew of safe, effective medications. If you have prehypertension (see the table on page 162), lifestyle changes may help you stave off hypertension for years to come. If you fall into the hypertension category, you'll probably need drugs to fight the disease, although the right habits can reduce the dose you need.

Here are the changes you can make:

- **Excess weight.** Lose it. Obesity triggers a cascade of changes in your body (as described in Chapter 2), eventually raising your blood pressure.

- **Diet.** High levels of salt can raise blood pressure, but the magnitude of the effect depends on the person—some people are more sensitive to the effects of salt than others. Your best chance to change your blood pressure through diet is by following the draconian DASH diet, which promotes eating fruits and vegetables and strictly limits salt, saturated fat, and alcohol. For more information, see *http://dashdiet.org*, or jump to page 194 to learn the basic principles of healthy eating.

- **Exercise.** Your body benefits from any physical activity, even if it's just a brisk walk once a day. But the best way to battle high blood pressure is to include three 30-minute sessions of moderate-intensity aerobic exercise every week.

- **Stress.** Everyone encounters stressful events that cause a temporary increase in blood pressure. (In fact, even the stress of having a highly credentialed medical professional measure your blood pressure can cause a 30-point rise. This transitory phenomenon is called *white coat syndrome*, after every doctor's favorite attire.) However, the effect of chronic, unrelenting stress is still unclear. To be on the safe side, take the time to relax, get proper sleep, and avoid situations of powerlessness (for example, working for an abusive boss).

The Circulatory System

Although your heart gets all the glory, it's nothing more than a marvelously efficient pump that sits at the center of a complex system—and that system has many equal players. Your heart works in conjunction with a network of muscle-lined blood vessels, a family of specialized blood cells, and several other key organs (like the oxygen-supplying lungs and the blood-filtering kidneys). In fact, in many ways your heart is the weak point in the system—a fatally limited component that causes catastrophe if it fails.

The heart's most important partner is the *circulatory system*—a set of roads, freeways, and back alleys that link your body together in a complex network that transports blood. The cells of your body get just about everything they need through the circulatory system. (And if your cells need to get rid of something, the circulatory system usually handles that, too.) In short, your circulatory system helps to transform your body from a collection of self-interested cells into a thriving, collaborative community.

The Contents of Blood

The traffic that flows along the passages of your circulatory system is *blood*. If you're an average person, you have about 6 quarts of blood perpetually circulating through your body. (In soft-drink terms, your body has 16 Coke cans' worth of blood in it.)

We usually think of blood as a runny, red substance. However, if you broke blood down to its basic components, you'd start with a straw-colored liquid called *plasma,* which consists of water, dissolved nutrients, and a few other ingredients you'll learn about in a moment. The red color of blood comes from the oxygen-carrying red blood cells you first met on page 136.

> **Note** Despite popular lore, blood is never blue (even when it's inside your body, and even in royalty). When your blood is fully stocked up with oxygen, it's bright red. Oxygen-depleted blood turns a purple-tinged, dark-red color. Due to the way your skin refracts light, your veins may look blue, but the blood inside is always some shade of red.

Your blood's complex cocktail of substances includes:

- **Oxygen and sugar.** Your circulatory system delivers the essential nutrients and fuel every cell in your body needs to produce energy.

- **Waste products.** Your circulatory system takes potentially poisonous substances from the cells that excrete them to their appropriate disposal site. For example, it carries carbon dioxide to your lungs and delivers other toxins to your kidneys and liver.

- **Hormones.** These signaling chemicals let one body part influence another in ways both subtle and profound, and they travel from source to destination through your blood. Often, your brain starts the cascade of hormone release. (For example, to trigger your *fight-or-flight response*, your body uses hormones that speed up your heart, dilate your pupils, and constrict your blood vessels. On a much more extended timescale, your body releases growth hormones that reshape your body and start the changes of puberty.)

- **Germ fighters.** Your blood is the battleground on which your immune system confronts bacteria, viruses, and tumors. (You'll learn about your body's endless war against disease in Chapter 9.)

- **Clotting compounds.** Your blood is precious stuff. Fortunately, it has its own loss-prevention system: *platelets* that plug holes in your skin, and a tough protein called *fibrin* that forms a reinforcing mesh around the platelets. Together, these compounds prevent your entire blood supply from draining out of a minor cut.

Veins and Arteries: Your Body's Plumbing

Blood travels through two types of passages. Oxygen-rich blood travels from your heart to the rest of your body through *arteries* (arteries move blood *a*way from your heart). Oxygen-depleted blood makes the return trip back to your heart in *veins*. Together, arteries and veins are called *blood vessels*, and their winding tubes stretch tens of thousands of miles.

Understanding Blood Types

You've probably heard that each person's blood can be classified according to *blood type*. The term refers to a relatively minor characteristic of blood—specifically, the types of proteins that stud the surface of your red blood cells. This detail is important because it influences how your immune system differentiates friendly blood cells from potential enemies. If your immune system discovers red blood cells that don't match its expectations, it's likely to destroy them.

Usually, there's little reason to spend time thinking about your blood type. But it becomes critically important in one situation—if you lose a lot of blood and need to be topped up with someone else's. In this situation, the donor's blood type needs to be compatible with yours, or your body will launch a dangerous, debilitating war against the new blood cells.

More rarely, blood-type issues can occur during pregnancy. In some cases, an unborn child inherits a blood type from the father that's incompatible with the mother's. This in itself doesn't pose a problem, but when the mother is exposed to this blood (typically during delivery), her immune system may begin building antibodies that can destroy the foreign blood cells. Because these antibodies linger in her blood forever—they're essentially a stockpile of on-reserve weapons—they can cause problems for future babies in future pregnancies. Fortunately, modern medicine deals painlessly with this issue: Early in pregnancy, mothers are given a blood test and, if necessary, a vaccine that prevents the production of these antibodies.

Arteries are more than just passive tubes. Healthy arteries are strong, flexible, and lined with muscle. As your heart pumps blood, your arteries expand and contract rhythmically to help move that blood along. Veins are more passive—they transport blood steadily rather than pumping it in time with each heartbeat. Tiny valves prevent blood from seeping backward, and normal muscular activity helps it flow forward (which is why getting up and walking around helps your circulation). *Varicose veins* occur when the valves in the veins weaken and blood becomes trapped. The condition is mostly harmless, but sometimes uncomfortable.

Note: Unlike arteries, which generally travel under large muscles and close to your bones, veins often run close to the surface of your skin and are easy to spot in many places, such as your wrists, inner arms, hands, feet, and chest.

To better understand how your arteries and veins work, it helps to take a trip through them yourself. You can start at the pump that keeps the traffic flowing—your heart.

1. Inject yourself into the right side of your heart (which is on the left side of the front-facing picture on this page). The blood that arrives here is the dark, purplish kind that has little oxygen left in it. With a forceful contraction, your heart squeezes itself together and propels your blood toward your lungs.

2. As your blood moves through the tiny blood vessels that line your lungs, it releases carbon dioxide and picks up a full cargo of oxygen. New blood keeps arriving, driving the oxygen-filled blood back to your heart.

3. When your blood reaches your heart a second time, it enters the left side (on the right side of the diagram). Your heart contracts again, pushing this oxygen-rich blood out through the *aorta*, which is the largest artery in your body.

4. As blood flows through your arteries, it moves through smaller and smaller passageways, eventually flowing into a microscopic network of *capillaries*, which are the tiniest blood vessels in your body. Nutrients and oxygen pass through the walls of these capillaries into neighboring cells.

5. Having completed its nutrient and oxygen delivery, your blood picks up waste products (carbon dioxide and so on) and is ready to return to the heart. Eventually, the capillaries join back together to form tiny veins. These tiny veins combine to create larger veins, and they eventually unite in the massive *vena cava* veins that lead back to the right side of your heart, and back to step 1.

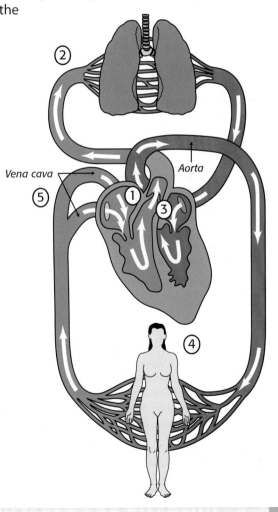

Vena cava

Aorta

Your blood makes this journey continuously, travelling from your heart to your lungs, back to your heart, through your body, and then back to your heart once more. Despite the winding route, each drop of blood makes the full circuit in less than 60 seconds.

Note Contrary to what you might expect, your veins don't reclaim all the fluid your arteries release. Instead, a small amount lingers in your cells, and is collected into another system of tiny passageways, collectively called the *lymph system*. The lymph system is primarily known as a part of your immune system. To fight infection, it filters lymph fluid through small, bean-shaped *lymph nodes* that are scattered throughout your body. The lymph nodes store heavy reserves of white blood cells that destroy any invaders. (You'll learn more about the workings of your immune system in Chapter 9.)

Heart Attacks

Cardiologists call it the "Hollywood heart attack." It's the one you see on television—a dramatic moment of chest-clutching pain that sends its victim crashing to the ground. Even though you've probably never experienced anything like it, you know instinctively how it feels—a crushing, squeezing pain, like having a portly man squat down on your chest.

Heart attacks can happen this way, but they usually don't. In the real world, the signs of a heart attack are both unpleasant and unclear. Unlike other medical emergencies, you can easily (and unintentionally) ignore its early stages. Grim statistics support this fact:

- The average heart attack sufferer waits nearly 2 hours before calling for help.

- Nearly half of all heart attacks are fatal.

- Women are more likely to ignore a heart attack because they don't tend to expect it, and they're more likely to have vague, irregular heart-attack symptoms.

Note Cardiologists call the first hour after the onset of a heart attack the *golden hour*. Get help in this window of time, and the odds are good that you can save your heart from serious damage. After that, you're much more likely to end up with a permanently crippled heart muscle, if you make it through at all.

Your Heart Attack

Hopefully, you'll never suffer a heart attack (at least not until you're good and ready, say with an age in the triple digits). But if your heart does run into unexpected trouble, it's essential that you recognize the symptoms and get help as quickly as possible.

An average heart attack may start modestly. Perhaps you've spent the day at the gym, and several hours later, you notice an unexpected tightness in your chest. Or you're driving home from an absolutely ordinary day at work, but feeling a nagging, uncomfortable shortness of breath. Whatever the problem, it's difficult to ignore—in fact, you feel thoroughly awful. You might also have aches, extreme fatigue, intermittent nausea, and sweating. But no matter what specific symptoms you experience, the one overriding characteristic of a heart attack is profound discomfort, not the clear-cut agony of a Hollywood heart attack.

> **Note** Don't make the mistake of assuming that discomfort needs to be continuous to be serious. Heart-attack symptoms may come and go intermittently, as the artery-obstructing clot breaks up slightly and reforms—but the danger is the same.

Up to Speed

A Heart Attack on the Inside

As you've learned, heart-attack symptoms can be vague and misleading. But inside your body, the situation is more straightforward.

The trouble develops over many years as tiny growths form on the inside walls of the arteries that supply the heart, causing the arteries to narrow significantly. These growths, called *plaque*, hold small deposits of fat. For reasons that aren't fully understood, some of these growths eventually become soft, inflamed, and ready to burst.

When one of these soggy problem spots tears open, it triggers an inflammatory reaction that causes your blood to form a *clot*. This clot seals the artery shut, in much the same way that it patches minor nicks and cuts in your skin. The result is that your heart is deprived of blood and starved of oxygen. The heart muscle begins to die.

Blood clots don't just trigger heart attacks. A blood clot that forms elsewhere in your body can break off and travel through your blood, eventually becoming lodged in a smaller passageway. Depending on where the clot is, it can cut off the blood flow to other organs and damage them. A clot can even block the arteries that feed your brain, triggering a sudden loss of brain function called a *stroke*.

During a heart attack, you may feel pain in a seemingly unrelated place, such as your back, neck, jaw, or arm (on either side). This phenomenon, called *referred pain,* happens when sensory messages from your heart get scrambled across nearby nerves, which are connected to a completely different body part. As a result, your brain gets the wrong message. (Referred pain is also responsible for the ache that appears to originate from a limb that's been amputated, and the "brain-freeze" feeling of drinking a 1-liter slush drink in 60 seconds.)

In the midst of a heart attack, people often make two mistakes:

- **They try to rest it off.** Although the discomfort of a heart attack doesn't go away, many people try to explain it as the fallout from a sudden flu, an overly enthusiastic workout, or a basket of bad clams. Eventually, it becomes obvious that these explanations don't fit—but not before valuable time has ticked away.

- **They go to the hospital without an ambulance.** Most people are far too embarrassed to call 911 if they're still conscious and able to move. Instead, they get a family member to drive them to the hospital. Or they call a taxi. Or they walk on their own. All the while, oxygen deprivation is slowly destroying their heart.

> **Note** For some reason, many people have the unspoken belief that 911 is the number to call when someone *else* is in trouble—usually, someone who's already incapacitated (say, by a gunshot wound, a massive injury, or a Hollywood heart attack). But if you experience the symptoms of a heart attack, you have two choices: Swallow your pride or be embarrassed to death.

If you want to survive a heart attack, the right response is clear: Don't worry about being alarmist. Don't hope that the problem clears up with a bit of rest or wishful thinking. Don't talk yourself into the average 110-minute waiting period. Instead, risk embarrassment and call an ambulance. As cardiologists say, time is muscle. The more minutes you waste, the more heart tissue you sacrifice—forever.

As you wait for the ambulance, it's a good idea to chew an aspirin tablet, which can thin your blood and reduce the chance of severe heart damage. (You need to chew the aspirin to allow it to enter your blood quickly. Otherwise, you won't see much action for hours.) However, avoid the aspirin-chewing trick if you suspect another problem (like internal bleeding), in which case a blood-thinning drug may do more harm than good.

When you get to the hospital, modern medicine has several nearly miraculous treatments that can save your life. Clot-busting drugs can dissolve blockages, and an emergency surgery technique, called *angioplasty*, can dislodge obstructions and open arteries using a tiny, inflatable balloon.

Heart Attack Prevention

With its complications and uncomfortably high odds of lasting heart damage and death, a heart attack clearly tops the list of Life-Changing Experiences Worth Avoiding. But considering that heart disease is the single most common cause of death (at least in the U.S.), is it really possible to avoid cardiac calamity?

Surprisingly, there's a lot you can do to stay out of harm's way. The blood-pressure advice you learned on page 163 (stay active, lose excess weight, and avoid stress) is a good start. But if these practices aren't doing enough in later life, it's critically important to step up your game. In other words, forget your good intentions—your doctor needs to check your *blood pressure* and *blood cholesterol*, and use the full arsenal of modern medicine if these numbers are dangerously high. (As you've already seen, high blood pressure damages the delicate inner lining of your arteries. Excess cholesterol becomes trapped in artery walls, kick-starting the inflammatory process that leads to potentially dangerous plaque.)

If you're not convinced, consider this startling statistic: a 50-year-old man who doesn't smoke or have diabetes, and who keeps his cholesterol and blood pressure in the recommended range, has just a 5-percent chance of a serious heart event over the next half-century. But if he violates any one of these conditions, his chances rise tenfold, hitting 50 percent. Unsurprisingly, the vast majority of 50-year-olds fall into the second, far riskier category. They remain apparently healthy even as their cholesterol level and blood pressure creep up, laying the groundwork for the Big One.

A Heart Attack on a Stick

There's one more critical step to prevent a heart attack: Stay away from cigarettes. Their effect on heart health rivals the destruction they wreak in your lungs. In fact, if you're a chronic smoker, you'll realize greater benefits from kicking the cigarette habit than you will from starting a hard-core workout regimen *and* adopting a strict health-food diet.

The way cigarettes work their damage isn't entirely clear, but smoking seems to doom your circulatory system in several ways, reducing the supply of oxygen to the heart, raising blood pressure, increasing the level of dangerous LDL cholesterol (page 50), and making blood more likely to clot. Altogether, it's a cocktail of heart trouble that's far more potent than a daily Big Mac.

If you're still not convinced, consider these grim facts:

- Heavy smokers have double the risk of stroke, and two to four times the risk of a heart attack.

- When a heart attack strikes, heavy smokers are nearly twice as likely to die.

- Even light smokers (those who smoke one to nine cigs a day) are at a third higher risk for a heart attack.

Cardio Exercise

In Chapter 3, you learned how strength training compels your body to rebuild itself, forging stronger, healthier, and more efficient muscles and bones. The perfect complement to strength training is *cardio exercise* (also known as *aerobic exercise*, for reasons discussed on page 73). A regimen of regular cardio exercise improves the function of your lungs, heart, and blood vessels. Over time, it gives you greater endurance for heart-pumping tasks like running, dancing, and swimming. It also reduces your risk of a whole medical manual of health problems, from diabetes to depression. It might even help you dodge a coronary catastrophe.

Getting cardio exercise is cheap and easy. You don't need special equipment or expert advice. You simply need to do something tiring (say, jumping up and down in one spot, or dancing around in your pajamas). Do it until you start breathing heavily, and then keep at it for at least 20 more minutes. This may not sound glamorous (in fact, if you pick the pajama option, it's downright embarrassing), but the science is sound.

The best approach to cardio exercise is to follow a regular regimen three times a week. But—surprisingly—research shows that even a *single* session each week has a positive effect on heart health. This finding kicks the legs out of your last good excuse to stay on the sofa.

Types of Cardio Exercise

If you get tired of having your friends and family laugh at your impromptu jumping jacks, you can try a more traditional form of aerobic exercise. Here are some of the most popular:

- **Running.** Requiring nothing more than time and a decent pair of running shoes, running is the most straightforward way to get your heart pumping. It's also a marvelously adaptable exercise you can tailor to any fitness level. To go easy on yourself, alternate between brisk walking and a light jog. For hard-core training, maintain a steady jog with bursts of flat-out sprinting.

- **Jumping rope.** Like running, rope skipping is a highly portable way to get intense aerobic exercise wherever you go. You can't slip an entire gym in your pocket, but there's always room for a jump rope.
- **Swimming.** If you have a pool handy, you can use swimming as a full-body, low-impact form of exercise. Because it's so gentle on the joints, many people like to alternate swimming with other types of cardio exercise to create a weekly workout program.

- **Bicycling.** Like running, cycling is an activity you can do either indoors (on a machine) or outside (on the street). As a side benefit, the effort of keeping yourself balanced on two wheels strengthens the muscles throughout your body.

- **Cross-country skiing.** It's one of the most intense forms of full-body aerobic exercise around. Cross-country skiing has obvious disadvantages—for instance, you need a snowy trail and a lot of equipment—and newbies will find it extremely challenging. But it makes an excellent excuse to book that Nordic vacation.

- **Step aerobics.** Any aerobic workout routine can deliver the goods, but step aerobics stands out from the late-night infomercial gimmicks. Two decades after its creation, the basic idea (performing choreographed movements using a raised platform) remains wildly popular. Done right, step aerobics has the perfect mix of low-impact exercise and intense effort. To get started, you can try an exercise DVD. Or better yet, join a class at your local fitness club or community center.

- **Gym equipment.** Purists might say that a treadmill gives you all the effort of running with none of the scenery, but exercise equipment is the perfect solution for many casual exercisers. And even if the convenience of treadmills, steppers, and elliptical trainers doesn't seduce you, you just might fall for the shiny electronic gadgetry (such as programmed courses and integrated heart-rate monitoring). If not, there's always the wall-mounted television.

Tip The experts agree—the best way to choose a cardio exercise is to forget about calories per minute and pick something you enjoy doing and can easily integrate into your daily routine. This gives you the best chance of maintaining a regular routine—and regularity is far more important than the type of exercise you choose.

Workout Intensity

To get the greatest benefit from aerobic exercise, it's important to pace yourself. Work too hard, and you'll exhaust your body and even risk injury. Take it too easy, and you'll waste valuable workout time with no obvious payoff.

If you want to get serious about optimizing your workouts, start by listening to your body's built-in exercise-intensity meter—your heart rate.

Maximum heart rate

To properly interpret your *exercise* heart rate, you first need to calculate your *maximum heart rate*. This is the highest speed your heart can reach and maintain during strenuous exercise.

The best way to determine your maximum heart rate is with a *cardiac stress test*, where your doctor hooks you up to an electrocardiograph machine and you run on a treadmill, all under the watchful eye of medical staff. If you've already had a stress test and know your maximum heart rate, you can skip ahead to the next section. Unfortunately, most people don't have this opportunity and have to make do with a less reliable, age-based formula.

The basic maximum heart-rate calculation starts with a number (220) and asks you to subtract your age. This is the equation that's used by heart-rate monitors and treadmills:

$$HR_{max} = 220 - age$$

This version of the formula is simple, straightforward, and about as reliable as a lifejacket made out of tissue paper. Fitness experts prefer the following tweaked version, which lowers the ceiling for maximum heart rate and reduces the effects of aging.

$$HR_{max} = 205.8 - (0.685 \times age)$$

For example, using this formula, the maximum heart rate for a 35-year-old is:

$$HR_{max} = 205.8 - (0.685 \times 35)$$

$$= 205.8 - 23.975$$

$$= 182$$

Of course, this is simply an average, and individuals vary. Your maximum heart rate could easily be 10 beats off the maximum that this formula predicts.

Target heart rate

Once you know your maximum heart rate (or at least have a good, working estimate), you can use it to shape the perfect workout. Here are the three basic principles to follow:

- By comparing your heart rate during exercise with your theoretical maximum rate, you can determine how hard you're working. A heart that's running at 90 percent of its maximum rate is working far harder than a heart that's pumping at 60 percent.

- A typical workout includes different levels of exertion. For example, your heart beats more slowly during a warm-up and cool-down than it does in the middle of your exercise routine.

- Depending on your exercise goals, you'll want to spend different amounts of time at different levels of exertion. For example, if your goal is to achieve basic heart health and burn fat, you can follow a less rigorous exercise program than that of an endurance-training athlete.

The following table outlines the different levels of exertion your heart can handle. For example, the "healthy-heart zone" is defined as the range between 50 percent and 60 percent of your maximum heart rate. If you're an average 35-year-old (with a maximum heart rate of 182), you're in the healthy-heart range when your pulse runs between 91 and 109 beats per minute.

Note Because the formula for finding your maximum heart rate isn't always precise, it's worth double-checking your aerobic zone using the talk test (see the far-right column in the table). Although the talk test seems somewhat unscientific, studies show that it's surprisingly accurate—in fact, it's roughly as reliable as the maximum heart rate calculation.

Aerobic Zone	Description	Heart-Rate Percentage	Comparable Talk Test
Healthy-heart zone	This is the light-exercise zone where you warm up. If you're making the abrupt transition from couch tuber to marathon runner, it's a good idea to stay in this zone for your first week.	50%–60%	Speaking is effortless, and you even have leftover lung capacity to sing a song.
Fitness zone	This is often called the "fat-burning zone" because it's perfect for casual exercisers who want to shed pounds.	60%–70%	You're able to carry on a conversation without much effort.
Aerobic zone	This zone steps up the intensity to challenge your heart and lungs. It's also called the "endurance zone" because it's the best place to be if you're preparing for an endurance event like long-distance running. Serious exercisers spend most of their time here.	70%–80%	You're able to speak in complete sentences, but not more, and prolonged talking is an effort.
Threshold zone	This zone is also known as the "performance-training zone." Athletes train at this level to increase their bodies' ability to process oxygen and fight fatigue.	80%–90%	Speaking is labored, and the most you can get out are short bursts of a few words at a time.
Redline zone	This is the most intense level of exercise your body can perform. Most people can stay in this zone only for a few minutes. To be safe, don't train in this zone unless you're in very good shape and have your doctor's clearance.	90%–100%	If you try to talk, you're limited to gasping single words.

Note Dizziness, lightheadedness, and pain are all clear signs of overexertion. If you experience any of these symptoms, stop or slow down.

Creating a personalized workout is like crafting a fine coffee. You mix different zones to get the perfect blend.

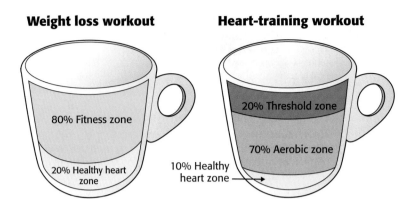

Weight loss workout

80% Fitness zone

20% Healthy heart zone

Heart-training workout

20% Threshold zone

70% Aerobic zone

10% Healthy heart zone

If weight loss is topmost on your mind, go heavy on the fitness zone, which maximizes fat burning without tiring you out. You can then stretch your workout to an hour for even better results. A typical weight-loss exercise mix might put 20 percent of your time in the healthy-heart zone (for your warm up), and 80 percent in the fitness zone.

If you're in decent shape and looking for the best boost for your heart, emphasize the aerobic zone. A typical mix puts 10 percent of your time in the healthy-heart zone, 70 percent in the aerobic zone, and 20 percent in the threshold zone (usually in bursts of 3 to 10 minutes at a time).

If you want to add high-intensity training in the redline zone, where your heart rate is near its maximum, do so with even shorter bursts of exercise of 30 seconds to 3 minutes. But be careful you don't overdo it. If you're an occasional exerciser, an overly ambitious cardio workout can leave you exhausted for the rest of the week.

Recovery heart rate

Earlier in this chapter (page 160), you learned that your resting heart rate provides a subtle clue about your body's overall fitness. The problem is that individual variation makes it difficult to identify anything but the extremely fit or unfit. There's another metric that tells you a bit more about the state of your heart—the *recovery heart rate*, which measures how quickly your heart calms down and returns to its normal rate after vigorous exercise.

There are many different recovery-rate tests. One is to end your workout with a 5-minute cool-down and then measure your heart rate 15 seconds after you stop. If your heart rate still hangs over 120 beats per minute, there are several possible explanations—your workout was too strenuous, you didn't have a proper cool-down, or your overall fitness level is poor.

8 Your Digestive System

G lamorous it isn't. But stretching from your lips to your anus is a 30-foot-long, mucus-lined hose that converts a perfectly appetizing meal into the all-too-familiar brown sludge.

Talk about it in public, and you'll quickly lose your dinner companions. That's because few people—aside from late-night comics or a roomful of toddlers—want to know what really became of yesterday's lunch. Oh, we may open our mouths to toss in a tasty morsel, but when it comes to talking about what happens next, our lips are sealed. And we certainly don't want anyone in a white coat messing around in there either, whether it's a dentist poking in the front end or a colonoscopist heading up the back.

That's unfortunate, because a proper understanding of your body's food-processing system can save you a good deal of embarrassment. For example, knowing your plumbing can help you fight bad breath, disarm heartburn, and ensure that everything you eat keeps moving steadily along. Most important, it gives you valuable insight into the nutrients you need in your diet and the troublemaking substances you should avoid.

In this chapter, you'll travel down your long, snaking food canal. You'll watch as your digestive parts—mouth, stomach, intestines, and everything in between—grind food to a fine paste and mix it with a powerful assortment of food-dissolving substances in the never-ending chemistry experiment we call *digestion*. On the way, you'll learn some of the secrets to proper nutrition, disease avoidance, and long-term health.

In Your Mouth

Two scrambled eggs, two slices of toast, and a breakfast sausage. Depending on your perspective, that can add up to breakfast, or it can be the start of a slightly unsettling tour that stretches from your mouth to a porcelain bowl. In this chapter, you'll follow this sample meal on its journey through *you*.

Your breakfast begins its trip through your digestive system in the place where you first put it—your mouth. Here, you take the initial step in the digestive process by chewing your food. Chewing breaks your meal down into small pieces, providing more surface area for your digestive chemicals to do their work. At the same time, your mouth douses each mouthful of food with generous amounts of saliva, which starts breaking it down.

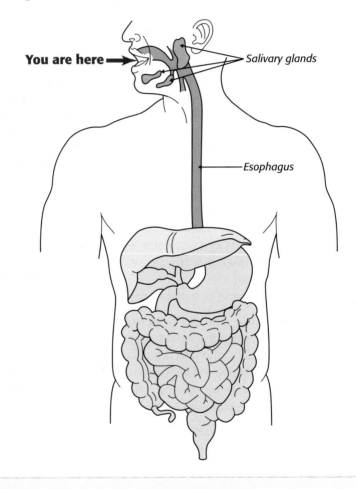

You are here ➡   *Salivary glands*

Esophagus

> **Note** If you're still feeling squeamish, consider this maxim: If you presume to consume, you can't protest to digest. In other words, if you're willing to put food in your mouth, be ready to face what your body does with it.

Interestingly, you have conscious control over just two parts of your digestive system—chewing, when food enters the system, and excreting, when leftover waste makes its exit. So try to enjoy both ends.

Frequently Asked Question

Do You Need Thirty Chews?

Few of us count the number of times we chew before swallowing a morsel of food (and those who do aren't much fun at dinner parties). Still, the age-old question remains: Can swallowing half-chewed chunks of food harm your digestive system?

The answer, it turns out, is fairly sensible. You should chew your food until it becomes an easily swallowable, broken-down paste. (One rule of thumb is that the texture shouldn't be recognizable. For example, if your potatoes still have skin on them and your broccoli still has a stalk attached, they're not ready to swallow.)

Basic chewing has several benefits. First, it triggers activity in other parts of your digestive system, like your stomach, preparing them for the work ahead. It also helps the rest of your digestion work more smoothly, because your digestive system doesn't need to struggle to extract the nutrients from large, tough chunks of food. Finally, chewing forces you to eat more slowly, which can reduce the chance that you'll choke or overeat, and improves your mealtime enjoyment (and dinnertime popularity). However, there's no reason to get obsessive about it. For example, the infamous motherly advice to count 30 chews will turn all but the most overcooked steak into something not unlike a wad of pulp from a paper mill—and who wants to swallow that?

Incidentally, extreme chewing was the foundation of a wildly popular Victorian-era diet. Horace Fletcher (also known as "The Great Masticator") became a millionaire by championing obsessive chewing as a way to avoid constipation and other digestive ills. His strict advice was to chew each mouthful more than 30 times over the course of a half minute, and then tilt your head back to let the result trickle down your throat. You were to spit out any remaining chunks. The side effects included long meal times, dramatic weight loss, and a sore jaw.

Your Teeth

If you were asked to pinpoint the hardest substance in your body, you might pick your bones. But the honor actually belongs to a far sturdier substance: the *enamel* that protects your teeth.

Your teeth have three distinct layers. The hardest material is the mineral-rich enamel on the outside, which absorbs daily wear. However, enamel is also brittle—bend it just a little and it snaps. To improve its resiliency, your teeth back this enamel with a thicker substance: a layer of yellowish *dentin*. Dentin isn't quite as strong as enamel, but it's less likely to crack under pressure. Finally, underneath the dentin sits the living pulp of your tooth, a soft tissue that nourishes the dentin and contains blood vessels and nerve endings that connect to the rest of your body.

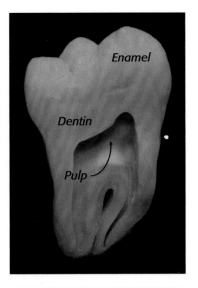

Enamel

Dentin

Pulp

Note If the pulp in a tooth becomes damaged—if a stray hockey puck lands in your mouth, for example—its living tissue may die, darkening your tooth. At this point, it's time to remove the remaining pulp and stuff the once-living tooth with a rubber-like filling. This operation is the infamous *root canal*.

Your mouth contains more than teeth and tongue. An impressive population of bacteria lives on all the surfaces in your mouth. Some of these bacteria feed on sugar, coating your teeth with an acidic byproduct called *plaque*. Left to its own devices, plaque can dissolve the tough enamel that protects your teeth and cause a *cavity*. Studies of ancient human skeletons suggest that cavities are far more common in modern times, no doubt due to our newfound access to sweet substances.

Tip In about 48 hours, plaque hardens into concrete-like *tartar* that only your dentist can remove. So don't skip a day of brushing.

Mouth Maintenance

Different animals have very different ways of producing and maintaining their teeth. Sharks grow a limitless supply and never need to see a dentist.

Elephants get a lifetime allotment of just 24 teeth and use no more than six at a time. As an elephant grinds its way through its tough, plant-based diet, its front teeth wear down and eventually fall out, making room for its back teeth to move forward as replacements. When an old elephant goes through all of its teeth, it starves to death.

Humans aren't quite as bad off—if we lose our limited set of 32 teeth, we can survive on protein shakes and peach ice cream. But if you want to bite your way through chocolate bars and ciabatta bread well into your nineties, you need to make sure your dental hardware doesn't end up as used and abused as a mouthful of elephant teeth.

Here's what you need to know to maintain healthy teeth:

- **Brushing.** It's not necessary (or helpful) to attack your teeth with sand-blasting force. Instead, a gentle 3- to 4-minute brushing does the trick. (Most people think they brush even longer, but the average brushing session lasts just 60 seconds.) Twice a day is the official tooth-brushing recommendation, but dentists really want you to clean your teeth after every meal.

Tip Modern research suggests you wait 20 minutes after a meal before you brush. Immediate brushing attacks your teeth when they're at their softest—weakened by the acids in your food.

- **Flossing.** Studies suggest that proper flossing might do more for your teeth than brushing. Flossing once a day won't bankrupt your dentist, but it will remove tiny particles of food between your teeth and reduce the sticky buildup of plaque. To get the most out of flossing, however, you need impeccable technique. Particularly important is flossing gently under your gumline, as shown in the picture below. For a step-by-step walkthrough, point your Web browser to *http://tinyurl.com/cltjjy*.

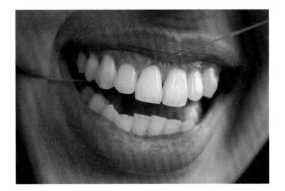

Tip Too pressed for time to clean your teeth with a strip of floss? As sardonic dentists often remark, you need to floss only the teeth you want to keep.

- **Whitening.** Stained, discolored teeth get little love. But recently, tooth fashion has switched from cloud white to glistening-Chiclets white, and the effect can be as glaringly unnatural as a nose job on Cyrano de Bergerac. If you really want a brighter shade of white, skip whitening toothpastes—most of them simply include abrasives that can grind away some surface marks. However, home whitening kits and custom dental appliances can produce better results, so talk to your dentist about what sort of whitening product would be most effective for you. This is particularly important if you've had any serious dental work.

- **Gum disease.** The real danger of poor dental hygiene isn't cavities (which are usually easy to patch), but *gum disease*. Gum disease occurs when the same bacteria that attack your teeth slip under your gumline, damaging your gums (in which case it's called *gingivitis*) or into your teeth's supporting tissues and bone (in which case it's called *periodontitis*). The former can make your gums swell and bleed, and can trigger bad breath. The latter can destabilize your teeth and lead to a set of dentures.

- **Fluoride.** Conspiracy theorists aside, fluoride plays an important role in strengthening tooth enamel, especially early in life. Countless studies have nearly always agreed about fluoride's cavity-preventive abilities and its lack of side effects, which is why it's so often included in municipal water supplies or (in many non–English-speaking countries) added to table salt. The only catch is that too much fluoride can stain the teeth of young children, but that won't happen unless they abuse fluoridated mouthwashes or eat entire tubes of toothpaste. (Incidentally, in the 1950s, far-right activists opposed fluoridation and vaccination, believing both were part of a shadowy conspiracy to impose a communist regime on America. So consider yourself forewarned.)

Note There's a more controversial ingredient in some toothpastes: *triclosan*, an antibiotic that coats the teeth. Studies confirm that it's an effective tool against the bacterial marauders that live in your mouth and cause plaque. However, some health experts worry that it could lead to bacterial resistance (page 218)—in other words, the presence of triclosan could encourage harmful bacteria to evolve into a super-species that's immune to the usual antibiotic weaponry. If you're looking to give your teeth an antibiotic boost—or if you just want to avoid this high-powered ingredient—check the label. Colgate Total is the best-known toothpaste to include triclosan.

Bad Breath

Basic dental hygiene removes particles of trapped food that can create an unpleasant mouth odor as the food decays. But a variety of other things can contribute to bad breath, such as:

- **Dry mouth.** Without the cleansing power of saliva, dead cells build up and decay in your mouth. This is the source of the phenomenon called *morning breath*, and it's particularly bad if you sleep with your mouth dangling open.

- **Digested food.** Certain foods, like garlic, have volatile oils that can stink up your airways. There's no way to rid yourself of these odors, because the odor begins after you absorb and process the food—and it actually seeps out of your *lungs*. Fortunately, the scent should die down in 24 hours. In the meantime, you can try to mask it by chewing on a clove, some mint, or a sprig of parsely (and hope that this combination doesn't create a still more objectionable smell).

- **Dental problems.** When bacteria works itself into places it shouldn't be—such as the pockets between your teeth and gums—it's impossible to remove on your own. The problem usually begins with poor dental hygiene, and you can only fix it with a trip to the dentist.

- **Diseases.** Certain medical conditions can produce strange or offensive smells. For example, untreated diabetes can cause a fruity smell. Kidney failure can cause an ammonia-like smell. If you suddenly develop a new and unpleasant mouth odor, check it out with your doctor.

Saliva

Saliva is a watery, frothy substance that your mouth manufactures continuously. It trickles out through glands scattered around your mouth, cheeks, and throat, but most of it seeps up from under your tongue. You produce a small milk-carton's worth (1 liter) of the stuff every day, most obviously when you eat, and hardly at all when you sleep.

Saliva cleans your teeth, lubricates your mouth, and protects the tender tissues inside. It also dissolves the substances in your food so they can reach your taste buds (page 128). In fact, without saliva, your favorite meal would be as appetizing as a stick of chalk.

Saliva also contains enzymes that start breaking down the long chains of complex carbohydrates in your food. (*Enzymes* are special compounds your body builds and then uses to carry out complex chemical reactions, like digestion.) For example, in the breakfast meal you're chewing right now, enzymes split some of the starches in your toast. It's all a bit of a preview to the heavy-duty digestion that takes place lower down the digestive tract.

Saliva also moistens your food so your teeth can compact it into a small, soggy ball that's ready to shoot down your throat and into your stomach.

> **Note** Although saliva includes its own natural antibacterial agents, opinions differ about whether the distinctly icky practice of licking wounds is healthy or dangerous. One thing is certain: Putting your tongue to a cut introduces not just antibacterial substances, but a huge family of mouth-dwelling bacteria, too—some of which can cause real trouble if they make their way deeper into your body.

Once your teeth and salivary glands have done their work, it's time to swallow hard and move on. Your meal has now entered the winding passages of your digestive tract, where it will remain for the rest of its journey.

Fun Facts

The Versatility of Spit

Many animals use their saliva for tasks other than eating. Nasty snakes poison their saliva with venom and inject it into their prey. Caterpillars use it to craft strands of silk that let them dangle from a leaf (and later, build a cocoon). Resourceful birds use it to construct tiny nests. (These gummy nests are used to create the signature texture of Chinese *bird's nest soup*. You can read all about this pricey, spit-soaked delicacy at *http://en.wikipedia.org/wiki/Bird's_nest_soup*.)

But if you think the animal kingdom has the last word in salivary creativity, consider this: For decades, art curators have cleaned ancient paintings with cotton balls soaked in spit (presumably their own). It seems that the same enzyme that breaks down the carbohydrates in your dinner can clean the grime off a picture. So the next time you're reflecting on a fine Renoir, don't be afraid to drool.

Your Stomach

When it comes to digestion, the first organ most people think of is the stomach—a pink and stretchy, j-shaped sack of muscle.

Contrary to what you might expect, your stomach isn't the end of the digestive journey. Instead, it's simply a short-term stopover. Your stomach's primary purpose is to warehouse food until it's ready to be passed along. At the same time, your stomach continues the work your mouth started, further breaking down your food physically and chemically.

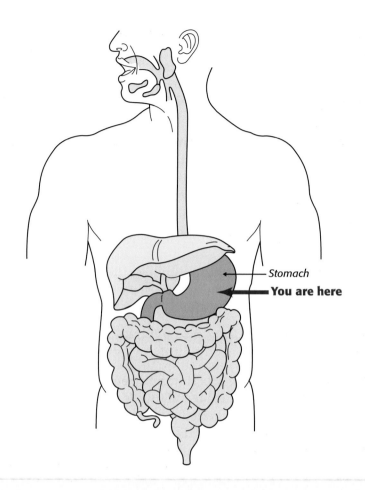

Stomach

You are here

> **Note** Your stomach is surprisingly inessential. Lose your stomach, and the rest of your digestive system picks up the slack, doing everything your stomach would do except for absorbing vitamin B12. In fact, people who have their stomachs removed (for example, sufferers of severe stomach cancer) can live perfectly well with a modified eating schedule that emphasizes smaller, more frequent portions and periodic B12 injections.

The Gastric Storage Tank

Your stomach is a temporary storage tank. Empty, it holds less than a child's juice box does. Swollen with food, it expands a staggering 30 to 50 times—large enough to accommodate nearly a gallon of food and drink.

Your stomach is also a muscle—one that few of us have trouble exercising. It expands and contracts continually, kneading, twisting, compressing, and mixing your food with powerful gastric juices that help digest it. The more food you put in your stomach, the more vigorous the mixing. When your stomach is empty, you might hear the noisy rumbles that biologists call *borborygmus* (pronounced "bore-bo-rig-mus"), but you call growling.

As your stomach churns your food, the meal gradually takes on the consistency of a creamy paste. Roughly three times a minute, your stomach squirts out a small eyedropper's worth of this paste into your small intestine (which is the next stage in the digestive journey). In this way, your stomach slowly works through your breakfast, preparing it for further digestion, being careful not to hurry the job and overwhelm your intestines.

Note Your stomach passes along almost all the food it receives. However, some substances can dissolve through the thick coating of mucus that lines your stomach and enter your blood. Examples include alcohol and certain drugs, like aspirin.

It usually takes 2 to 4 hours for your stomach to empty itself. Fluids and carbohydrates pass through it quite quickly, while protein takes longer, and fat forms an oily layer that's digested still more slowly. Large, fatty meals can linger for 6 hours or more. In the sample breakfast meal used here, the last holdout is the sausage, which supplies half of its calories from fat. For that reason, the sausage is also the most likely part of your meal to return as heartburn (see the box on the following page).

Tip Ideally, you'll eat food that won't race through your system *or* overstay its welcome. If you eat meals that have a dash of unsaturated fat and a good dose of fiber, you'll process them steadily but gradually, and you'll feel full longer. You'll also avoid the sudden sugar rush and insulin release that raw carbohydrates cause—a process that can, over the years, encourage diabetes.

Heartburn

While your stomach churns, two valves keep its contents tightly contained. This is important, because the mixture of food and gastric juices is highly acidic and decidedly unwelcome in other parts of your body. If the topmost valve fails and the acidic mixture escapes up your esophagus, the result is *heartburn*.

As you already know, heartburn has nothing to do with your heart, although the burning chest pain can mimic heart trouble. To avoid heartburn, try these tips:

- **Elevate yourself.** Sit up after a meal, and use pillows or a wedge to prop your upper body while you sleep. This enlists the aid of gravity. (Also, it's a good idea to avoid eating before you plan to lie down, such as in the 2 hours before bed.)
- **Eat small portions.** When your stomach is swollen with food, it's easier for the acidic mixture to burst loose.
- **Avoid trigger foods.** Most heartburn sufferers can pinpoint problem foods that cause excess acid production. Your list won't be the same as someone else's, and potential problem foods—such as spicy meals, acidic fruit drinks, and fizzy soda pop—may be either harmless or exquisitely painful once they're in your stomach.
- **Don't squeeze.** The stomach is a soft pouch. Restrictive clothing, a tight belt, or a hefty layer of subcutaneous fat (page 51) can put pressure on your stomach, encouraging it to squeeze open like a tube of toothpaste.

If you suffer from an occasional bout of heartburn, your best bet is to treat it with an over-the-counter antacid. Avoid milk—although it can temporarily soothe the stomach, the proteins it contains will soon stimulate increased acid production and possibly make your heartburn worse.

Finally, don't ignore persistent heartburn. If heartburn strikes two or three times a week for more than 4 weeks, it's time to bring in a doctor to check for more serious chronic problems. And if you have heartburn that gets worse before meals and fades away as you eat, it may be the sign of an ulcer (a tiny sore in the lining of your stomach), which doctors can often treat with a simple course of antibiotics.

Stomach Acid

Your stomach breaks down food using a solution of *hydrochloric acid*, a strong chemical used in industrial settings to strip rust off steel, and whose corrosive power rivals that of battery acid. On an average day, your stomach secretes about 2 or 3 liters of the stuff.

Stomach acid serves two purposes. First, it kills a range of bacteria that could pose a problem to your body farther down the digestive pipeline. Second, it activates other digestive enzymes that begin breaking down protein. In the sample breakfast meal, this is when your body begins working on the protein in your eggs and sausage.

If stomach acid escapes your stomach, it can cause heartburn. But your stomach itself is mostly immune to the effects of the acid because it's coated with a thick layer of mucus (which prevents it from digesting itself). Also, your stomach rapidly sloughs off and replaces the cells that lie underneath this mucus, giving you an entirely new stomach lining every few days.

The Practical Side of Body Science

Fasting and Detoxifying

Fasting is the age-old practice of limiting food and drink, often as part of religious festivals. Some fasts restrict all food, while others allow a stripped-down diet. Fasts may also go hand-in-hand with non-food restrictions, such as religious rules forbidding fighting, lying, and sex. (Fasts are notorious for lumping sin and pleasure together into one giant category of forbidden pastimes.)

The benefits of fasting aren't digestive. Supporters point out how practicing self-restraint (and enduring a little borborygmus—see page 188) develops inner will. They're less likely to point out the way fasting *increases* carnal pleasures post-fast. Much in the same way that you feel good when you stop striking your head against a tree, the end of a fast brings a heightened appreciation of everything you temporarily sacrificed.

More controversially, fasts are sometimes studied as a way to improve health. Some studies suggest that occasional fasting or lifelong calorie restriction can boost life expectancy. One possible reason for this phenomenon (if it actually exists) is that gentle stress may prompt the body to fire up certain beneficial repair processes. Or it may simply be that less food means less of all the ills of the modern diet—from excess sugar to runaway fat.

The score for so-called detoxifying diets is far less promising. Promoters suggest that extreme fasting, bizarre diet restrictions, or colon "cleansing" (which, happily, this chapter will not attempt to illustrate) can purge toxins from your body. The idea is alluring—after all, who wouldn't like to atone for a lifetime of dietary sin and return the body to a pristine, unpolluted state? However, the science is about as solid as a bowl of low-calorie Jell-O.

Your Small Intestine

Several hours after it starts its journey, your breakfast meets one of the wonders of the human body—the narrow, winding tubing of your small intestine. Measuring some 21 feet in length (but only about an inch wide), your small intestine is the longest part of your digestive tract. It's folded, wrapped, and wound around itself to fit in the tight confines of your abdomen.

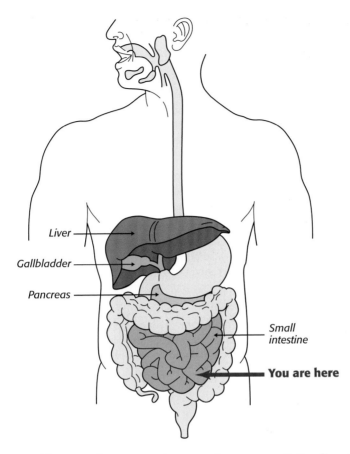

Liver

Gallbladder

Pancreas

Small
intestine

You are here

Small, wavelike muscle contractions push your partially digested food through your small intestine. These movements, called *peristalsis*, occur 8 to 14 times each minute, ceaselessly moving and mixing your food. The entire trip from one end of your small intestine to the other takes a relatively brisk 3 or 4 hours.

The Great Absorber

Your small intestine has two important responsibilities. First, it finishes digesting your meal, further breaking down its proteins, fats, and starches into simpler compounds. To perform this task, it gets help from several accessory organs, including your pancreas, liver, and gallbladder. Second, your small intestine absorbs your meal's final, fully digested nutrients, pulling them out of the pasty digestive solution and passing them directly into your bloodstream, making them available to the rest of your body.

Using Intestines from the Animal Kingdom

Your intestines are quite important, and you'd be ill advised to part with a single foot of the stretchy tubing. However, it just so happens that the resilient tissue that lines your gut lends itself to a host of arts and crafts. Humans have capitalized on this with the help of other animals. In fact, we're downright notorious in the animal kingdom for putting the intestines of other species to work in a variety of creative ways. Here are some examples:

- **Cheese-making.** Cow, sheep, and goat guts contain *rennet*, an important additive in the cheese-making process. Presumably, ancient man discovered this while making cheese in a convenient sack—the stomach of a dead animal.

- **Music-playing.** For centuries, craftsman fitted violins and other stringed instruments with tough fibers made from animal intestines. Although silk, nylon, and steel are more common today, some top-caliber musicians insist that nothing can match the sound of fresh sheep gut.

- **Food.** Natural sausage casings (the thin, plasticky substance that wraps your breakfast sausage) use animal gut from a pig, cow, or sheep.

- **Sex.** The world's oldest known condoms (dating back to about 1640) were made from sheep intestines. They were quite expensive, which probably accounts for the roaring trade in washed, second-hand condoms that prevailed at the time.

The easiest way to understand what's going on in your small intestine is to consider the different types of food it processes:

- **Carbohydrates.** Several hours ago, you began to digest your breakfast toast in the relatively easygoing environment of your mouth. Now, deep in your abdomen, the process continues with a new series of more powerful enzymes that shatter the remnants of your toast into simple sugars. Your body secretes these enzymes from your *pancreas*—a plain-looking slab of an organ that squirts digestive juices into your small intestine.

- **Proteins.** More recently, your stomach started working on the proteins in your eggs and sausage. Your small intestine finishes the job, again with the help of enzymes from your pancreas. The end results are *amino acids*—fundamental building blocks your body uses to assemble hundreds of thousands of different biological compounds.

> **Note** With the pancreas's high-powered digestive abilities, you might wonder why it doesn't eat itself. The trick is that the pancreas releases harmless, inactive enzymes. These enzymes switch themselves on when they meet up with the strong acids in the partly digested food in your intestine.

- **Fats.** Your body has to break these down into *fatty acids*. Once again, the pancreas secretes the enzymes your small intestine needs to do the job. However, before these enzymes can get to work, your body needs a way to break the big, greasy globules of fat into tiny droplets, in much the same way that dish detergent dissolves the oil from last night's deep-fried chicken. Two organs solve the problem. First, the liver—a multifaceted organ whose main responsibility is filtering blood—creates *bile* that does the trick. Second, the *gallbladder*—a kiwi-sized organ that looks like an unremarkable green pouch—stores and concentrates this bile between meals.

Once your body breaks down these nutrients, they seep through your thin intestinal walls, along with various vitamins and minerals. To make this process easier, thin folds lined with tiny hairs cover your small intestine. These details help increase your small intestine's surface area to promote nutrient absorption.

The Practical Side of Body Science

Purposefully Indigestible

Depending on your meal, your small intestine may contain quite a bit of undigested food. (This isn't the case with the simple breakfast example because it's short on plant matter and other sources of fiber.) In modest quantities, this undigested food benefits your digestive system and eases the passage of your meal as it scrapes through your narrow intestinal passageways. We call it fiber. One example of indigestible food is *cellulose*, a compound that helps form the structure of green plants.

In the dieting world, there's a completely different class of indigestible food—substances that masquerade as the sugar and fats our mouths expect, while dodging the absorption step in the small intestine. One example, *sucralose* (which is known commercially as Splenda), is a subtly modified version of sugar that triggers the sweet taste buds on your tongue, but can't be broken down by the carbohydrate-processing enzymes in your body. Another example is *olestra*, an altered fat molecule that has the same mouthfeel as fat but passes unhindered through your small intestine. (Eat olestra in great quantities, and you'll have a significant amount of unneeded matter moving through your system, potentially leading to the abdominal cramping and loose stools mentioned in the package warning label.) These two indigestible foods are examples of food science at its creative best—and potential health concerns.

The key concern for most sugar and fat replacements is not toxic side effects, but the way they allow non-foods with no nutritional value to take up valuable stomach space. Dieters caught up in the excitement of eating without weight gain may forget that their calorie-free potato chips are displacing real foods, and in the process robbing their bodies of the vitamins, minerals, and other nutrients they need.

Diet Advice from Your Small Intestine

One of your digestive system's limitations is that the body parts in charge of picking, identifying, and enjoying what you eat (your brain and your mouth) sit a great distance away from the parts that actually process and absorb your meal (mainly, your small intestine). So it's no wonder that this system so often slips out of sync, sending you hurtling straight into dietary trouble.

The real problem is that your brain and mouth follow a somewhat outdated ingredient list. They're always searching for the immediate gratification of sweet, calorie-dense foods. The rest of your digestive system simply processes whatever it gets. In 100,000 years of human evolution, it never occurred to anyone that people might somehow be able to consume vastly more food than they need.

The only antidote to the rampant abuses of modern eating is to simplify your diet with a few common-sense principles. And now that your small intestine has broken your breakfast down into its essential parts—protein, fat, and sugar—it's easier to pick out these principles, and to separate what your body eats from what it needs.

Here's what your small intestine would ask for if it had a voice:

- **Natural, unprocessed foods.** Industrial processing—the sort of thing that changes a box of ordinary rice into a package of breakfast cereal—amounts to pre-digestion. It takes responsibility for breaking down foods away from your small intestine, and it can eliminate trace elements of hundreds of different nutrients—all for a product that tastes like flavored packing material.

- **Plant-based foods, with small quantities of meat.** Unless you're a weight-training athlete, your body can get all the protein it needs from two servings of meat per day. (That's a piece of chicken, beef, or fish that's the size of the palm of your hand, without the fingers.) Dieticians often suggest treating meat as a condiment—in other words, as something you add to flavor a nutrient-rich plate of vegetables.

- **A rich variety.** Rather than obsess about the nutritional merits of squash versus sweet potatoes, strive to incorporate a range of healthy food into your diet. Highly varied dining offers another benefit: The sheer amount of healthy food tends to crowd out other, less desirable foods.

- **More complex carbohydrates, less sugar.** Your mouth, stomach, and small intestine eventually break down all carbohydrates into sugar. The more refined the carbohydrate, the faster the conversion, and the quicker you absorb it. This is a problem, because your digestive system is all about pacing (as demonstrated by the careful, one-squirt-at-a-time food transmission from your stomach to your small intestine). Heavily refined foods leave your stomach more quickly, which reduces your body's ability to pace itself and leads to see-sawing levels of blood sugar that your liver must work hard to adjust. But if you fill up with complex carbohydrates like vegetables, whole wheat flour, and brown rice, you'll have a tankful of food that will fuel you with a slow, steady supply of sugar for hours to come.

- **Water.** If your food comes premixed with fluid, you need less saliva and gastric juice to create the creamy paste your digestive system expects. And although well-meaning nutritionists sometimes warn heavy drinkers (of water) that they can dilute their gastric acids at mealtime, the effect is minor and has little effect on the average stomach.

Your Large Intestine

The large intestine, most of which is called the *colon*, represents the last processing stage of the digestive pipeline. Once your small intestine finishes extracting the last traces of nutrition from your meal, it pushes the unrecognizable leftovers into this thick, 5-foot hose, which circles the outer edges of your small intestine.

The large intestine is far shorter than the small intestine (the designations "large" and "small" refer to width, not length). Despite the short route, the last dregs of your food stay here longer than anywhere else in your digestive system. On average, the final sludge takes 10 to 12 hours to push its way through your large intestine. Once it reaches the end, it may hang around for several more hours before you finally get around to disposing of it.

By this point, your breakfast is greatly concentrated, reduced to one-tenth or less of its original volume. The task of the large intestine is to absorb any remaining water and to temporarily store the final product.

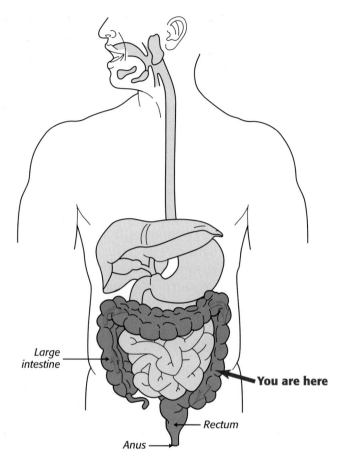

Large intestine

You are here

Rectum

Anus

Bacteria: Your Extended Family

Your large intestine is an active place. In fact, it's somewhat like a frat house full of partying bacteria. What attracts them is the all-night buffet of rich, organic waste that your body can't use.

Most of the bacteria living inside your large intestine are harmless cohabitants that share your body space without causing the slightest digestive disturbance. Some, however, are more irritating—they may release acids or a mixture of less-than-fragrant gasses that come out of your back end. Many species of bacteria are downright beneficial. As they feed on the indigestible carbohydrates you can't use, they synthesize valuable compounds like vitamin B and vitamin K, which your large intestine then absorbs. (In fact, if you could make your body bacteria-free, you'd need constant vitamin injections to stay alive.) Other species of bacteria are more defensive in nature—their presence acts as a check that makes it more difficult for new, harmful bacteria to set up shop in your insides.

Among these digestive dwellers is the infamous *E. coli* bacteria. We all carry our share of E. coli—in fact, it colonizes the pristine digestive tract of a new baby within 40 hours of birth, hitchhiking its way in on food, water, or caregivers. Although its presence sounds like a very bad thing indeed, it's rarely much to worry about, for two reasons. First, the variants of E. coli in your system are usually mild and relatively harmless, unlike the food poisoning kingpins that give E. coli its bad reputation. Second, your digestive system does its best to keep your bacterial family tightly locked up, using natural antibiotics and your immune system to hold it in check. The weak link in this system is you. If you fail to observe proper toilet hygiene—that means careful hand washing after a bathroom break—you're likely to spread E. coli and other bacteria from your hands to someone else's mouth.

Frequently Asked Question

Why Don't Mexicans Get Traveler's Diarrhea?

If you're a citizen of the world, you've probably met up with that uncomfortable phenomenon known as *traveler's diarrhea*—a short episode of diarrhea that strikes those brave enough to visit local places and enjoy local cuisine. The odd part is that traveler's diarrhea seems to affect only travelers. Native people can eat the same foods and emerge unscathed.

The first thing to understand is that traveler's diarrhea needs a certain level of sanitary sloppiness to occur. In particular, it only happens if there's some way for bacteria to pass from another person's (or an animal's) feces into your environment. For this reason, traveler's diarrhea happens much less often to visitors of most first-world countries. (Although Mexicans do occasionally get diarrhea when visiting the U.S., which they call "Washington's Revenge.")

However, this doesn't explain why locals have a much-reduced rate of diarrhea. The answer is that, because of near-continuous exposure, their digestive systems have gradually grown to recognize and tolerate strains of bacteria that other people can't handle. No one knows how long this immunity takes to develop or how long it holds up, but a study in Nepal found that American adults needed 7 years of local life to adjust, and they lost their tolerance after only a few months back home.

Interestingly, enterprising travelers can use one approach for instant immunity. If you're worried about E. coli (which is the most common culprit in Mexico), you can buy a vaccine called Dukoral that gives you temporary immunity. To get Dukoral, head to your local pharmacy or check with a travel clinic, which can also identify the gastrointestinal dangers in different parts of the world.

Balancing Your Bacteria

When it comes to your teeming family of bacteria, harmony is best. A careful balance of bacterial colonies prevents any one faction from causing trouble. It also prevents other invaders from establishing a toehold in your colon.

With that in mind, what can you do to encourage the right mix of bacteria? Some scientists believe it starts with birth. During the contractions of labor, it's common for a woman to pass a small amount of stool. Rather than being an unhygienic contaminant, this small parcel of feces might just be the mother's first gift to her newborn—a starter kit of healthy bacteria, ready to begin colonizing the world's newest colon.

After birth, it's up to you to maintain a happy bacterial family. The most significant factor is your diet, because the type of waste that passes into your large intestine determines the food supply that's available in your colon and, by extension, what species of bacteria will flourish. Once again, a diet rich in vegetables, fruit, and fiber presents the winning combination for intestinal health.

Many things can disturb the balance of bacteria in your large intestine, from a bacterial infection to a course of antibiotics. Over time, these colonies gradually reestablish themselves, but in the short term, your lack of digestive power can cause diarrhea or leave you at greater risk of an opportunistic bacterial invader.

Note Some researchers believe that the bacteria in your digestive tract plays an important role in priming your body's immune system to distinguish friend from foe. Some worry that commonly prescribed antibiotics may help trigger allergies and asthma when they wipe out these bacterial communities.

What Are Probiotics and Prebiotics?

Probiotics are live microorganisms (bacteria or yeast) that are particularly well-suited to your digestive system. For example, *lactic acid bacteria* is a common probiotic that gives sourdough and yogurt their characteristic sour flavor. In your colon, lactic acid bacteria digest the sugar known as lactose and may even prevent inflammation and inhibit cancer.

However, there's a catch. As you've seen, your large intestine is quite far down in your digestive system. For a probiotic to make it to its new home, it needs to pass through the inhospitable acidic environment of your stomach. Pharmaceutical companies are experimenting with special coatings that help probiotics make the hazardous journey intact, but in the meantime it's hard to tell how effective probiotic-fortified foods really are.

Prebiotics are substances that your healthy, colon-dwelling bacteria like to munch on. Supply these bacteria with more prebiotics, and you can encourage a small population to grow. Prebiotics are naturally present in fruits and vegetables, but don't expect to find any in a box of macaroni and cheese.

The bottom line is that both probiotics and prebiotics are based on valid nutritional science that recognizes the value of good gut bacteria. But their success as products is less clear, and it's a good guess that you'll get more benefit from a diet that emphasizes fruits and vegetables than one that focuses on convenience foods and nutritionally fortified drinks, no matter what miraculous new additives manufacturers toss in.

The Embarrassing Exit

The most frequent movements in your large intestine are slow, wavelike contractions that move food along its 5-foot length every 30 minutes or so. More powerful movements occur three or four times a day, and force the contents of the large intestine toward your temporary stool-holding tank, the *rectum*. Typically, these movements occur during or shortly after a meal.

Once your stool is in your rectum, conscious control returns to you. At a suitable time, you relax the flat shelf of muscle that keeps your stool in place. The product that emerges consists of undigested food, millions of bacteria, and the mucus and cells shed from the lining of your digestive tract along the way.

Some people examine their fecal matter as though it were a pile of fortune-telling tea leaves. So-called holistic healers are notorious for declaring that any stool more offensive than a toilet bowl full of jasmine leaves heralds a serious digestive imbalance. But before you get obsessed, understand that there's a healthy range of variety in the color, smell, and consistency of normal stool.

Generally, the best sign of a good digestive system is a stool that's soft and easy to pass. The ideal stool is like a brown banana, with a consistency no firmer than soft toothpaste. Dark, hard stools have probably spent too long in transit, and signal a need for more fiber. To size up your stool, find it in the Bristol stool chart (shown on the next page), which is a well-known medical aid that classifies poop. If your stool falls into the Type 1 or Type 2 category, you're mildly constipated.

More important than aspiring for the ideal stool is knowing what's normal for you. A sudden, prolonged change in stool frequency or consistency can signal a potential medical problem.

Note: In an average person, the sample breakfast meal takes a leisurely 2 days to travel the full digestive circuit. However, transit times range from under a day to 3 or 4 days. If you really must know how long your meal takes to travel through your entire digestive system, boil and eat some beets. The unmistakable red color will herald their arrival at the other end.

Smoothing the Exit with Fiber

Your large intestine is remarkably efficient at sucking the water out of your waste. After all, water is essential for life, and your body isn't willing to sacrifice a single drop. This is why prolonged diarrhea can be a severe health risk, particularly for young children and people in developing countries. With diarrhea, your body races to expel potentially harmful food wastes in a bid to get rid of toxins or bacteria before your body absorbs them. A significant amount of water sloshes out with the hastily expelled food. If you don't replace this water, the result can be severe dehydration and death.

Bristol stool chart

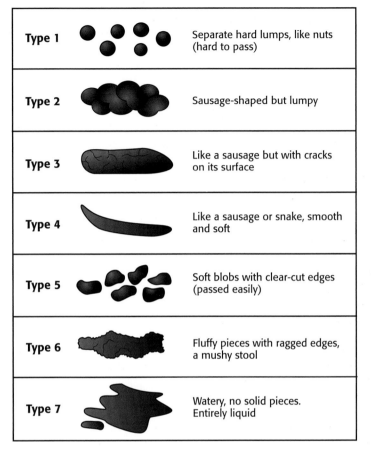

Type 1		Separate hard lumps, like nuts (hard to pass)
Type 2		Sausage-shaped but lumpy
Type 3		Like a sausage but with cracks on its surface
Type 4		Like a sausage or snake, smooth and soft
Type 5		Soft blobs with clear-cut edges (passed easily)
Type 6		Fluffy pieces with ragged edges, a mushy stool
Type 7		Watery, no solid pieces. Entirely liquid

A properly functioning large intestine is so efficient that it can remove too much water, leaving a dry, hard, and crumbly leftover that can't make the final stage of its digestive journey without causing some serious discomfort. This familiar phenomenon is called *constipation*. At its worst, it can lead to hours of pained straining and can inflame the veins in your back end (a condition called *hemorrhoids*).

Although constipation is a common complaint, it has a surprisingly simple solution. Minor lifestyle changes, such as getting proper exercise, eating modest portions, and always taking time to heed the call, can help. But the real miracle cure is *fiber*—indigestible matter that softens your stool and bulks it up, allowing it to pass through the final stage of the food canal quickly and easily.

There are actually two types of fiber. *Insoluble fiber* passes through your large intestine almost untouched. *Soluble fiber* absorbs water and develops a soft, gel-like texture. Both types help ease the passing of stool, but the processing of soluble fiber is linked to a host of health benefits that have nothing to do with toileting habits. For example, it may reduce blood sugar and cholesterol levels.

The average American gets just half the fiber doctors recommend for digestive health—and a typical teenager scores much lower, with only one-fifth of the minimum requirement. To boost your fiber intake, try adding these foods:

- Legumes, such as lentils, chickpeas, and other dry beans (lima beans, kidney beans, black beans, and so on)
- Many vegetables, such as broccoli, green beans, and carrots
- Many fruits, such as grapes and prunes
- Vegetable and fruit skins, such as potato and tomato skins
- Oats, barley, and quinoa (a wonder grain you can cook like rice)
- Nuts and seeds
- Whole-grain bread and brown rice

Note Some high-fiber foods (notably beans and other legumes) can cause embarrassing and uncomfortable intestinal gas. But don't let that put you off. Many people find that after several weeks of consuming a new high-fiber food, their body acclimates and no longer produces prodigious amounts of gas. Even if time doesn't solve your problem, you're likely to find that your reaction to high-fiber foods is highly idiosyncratic, and you may need to experiment to find the specific high-fiber foods that don't give you much gas.

You can also try a fiber supplement, which is the only form of laxative that's safe to use for prolonged periods of time. Fiber supplements aren't as effective as a diet that's naturally rich in fiber (and they lack the other nutrients that come with a fiber-rich diet), but they can help get irregular bowels back in line.

Note Despite what you may have heard, eating lots of fiber doesn't reduce the amount of vitamins you absorb from your food. If anything, it does the opposite.

How Many Glasses of Water a Day?

It's the question that everyone seems to ask. And if you follow the standard advice (drink 8 to 10 glasses of water every day), your next request will be for directions to the restroom. Because unless you're a strenuous exerciser or a desert dweller, you're unlikely to need that much water—and unless you're carrying a horse's bladder, you won't hold onto it for long.

No one's quite sure where the 8-to-10 glasses factoid started. However, medical professionals do agree on quite a few things about fluids:

- **Six glasses is usually enough.** If you must count, 6 glasses of water a day is probably a good rule of thumb (not a bare minimum). But the average person, doing gentle activity in a gentle climate, can probably get all the fluid they need from solid food alone (although it's not recommended).

- **Follow your thirst.** Your need for water varies greatly depending on your activity level. Fortunately, your body is surprisingly good at telling you when to drink. And the idea that we're chronically (and unknowingly) dehydrated is little more than science fiction.

- **Don't fear coffee and tea.** Despite the diuretic properties of caffeine, you'll still retain a large amount of the fluid in every cup—and even more if you're a regular drinker of caffeinated beverages.

- **Dehydration may worsen constipation.** If you're straining to pass stool, you might benefit from increasing your water intake a bit. However, results vary, and a more likely cause of constipation is inadequate fiber in your diet (page 200).

So why are we so easily misled by drinking myths that don't hold water? Quite simply, in the era of modern science, we're used to hearing (and accepting) startling facts. But when it comes to water, medical research is in an unusual position: proving that our common sense was right all along.

9 Your Immune System

By this point, you've probably realized the sorry truth. Your body is a pastiche, made up of both the amazing and the amazing-it-works-at-all. On one hand, your body features some of the most intricate, finely tuned biological machinery in the natural universe. On the other, it stuffs the gaps with quirks, messy compromises, and more than a few embarrassing shortcuts. It's a bit like a high-tech supercomputer that's patched together with duct tape.

Your immune system is no different. You probably think you have a good, basic idea of what it does. Perhaps you imagine something like a crack army of killer cells that hunts down and annihilates disease-causing agents. While this is true to a point, it glosses over several biology Ph.D.s' worth of details. The reality is that your body has a dizzying array of overlapping defense mechanisms that recruit dozens of different types of specialized body cells and pursue dramatically different strategies. The reason for this complexity is simple—in a battle for your life, it pays to use every weapon at your disposal.

You could easily spend your life studying the human body's defenses. However, unless you're hoping to impress a microbiologist on a first date, you don't need to. Instead, you need to know the basics—how your body defends itself, how you can keep your immune system strong, and how you can close some of its weak links before microscopic marauders exploit them to make you sick.

In this chapter, you'll cover these essentials. Along the way, you'll look at your body through the eyes of its enemies, from potentially dangerous bacteria like *E. coli* to the resilient viruses that cause the common cold. You'll even consider the problems that occur when your immune system runs off the rails and attacks *you*, triggering disorders from allergies to asthma.

Self-Defense

Your body is a comfortable home, but you aren't the only one taking advantage of its cozy hospitality. Over the course of an average day, armies of bacteria, viruses, fungi, and microscopic parasites swarm your body. They're hunting for a weak link in your body's defenses—and unfortunately for you, there are plenty.

The collective name for all these attackers is *pathogens*. They can crowd into your blood through fresh cuts, pass into your lungs as part of the air you breathe, and seep into your digestive system through the food you eat. When they do, they'll quickly come up against the aggressive and slightly paranoid forces of your immune system. Most will die a quick, unremarkable death. But some will stick around just long enough to cause some serious side effects.

Before you consider these troublemakers, you need to take some time to understand the three layers your body uses to defend itself. The following sections give you a quick tour.

The First Front: Physical Barriers

Your body wouldn't stand a chance of surviving in the hostile world if it weren't for the physical barriers that wrap its gentler bits. In Chapter 1, you learned about skin, which deters pathogens with a tough, lifeless outer layer of cells and a coating of antibacterial oil (page 21). Similarly, your body protects the lining of your respiratory system with mucus-covered membranes that trap germs, and it safeguards your digestive tract with a reservoir of stomach acid.

The Second Front: Inflammation

As important as your body's physical barriers are, they can't repel all invaders. Minor cuts in your skin open a brief, tantalizing entryway to your body's nutrient-rich interior. Tiny germs can eventually make their way through the thickest mucous coating. And while the acid in your stomach is strong enough to pickle steel, crafty microbes can survive or slip through it to your much more hospitable intestines. (For example, *H. pylori*, the culprit behind most heartburn-causing stomach ulcers, secretes an acid-neutralizing enzyme that protects the microbe until it gets a chance to worm its way into your stomach walls.)

When a pathogen breaches your body's first line of defense, the lowly foot soldiers of your immune system meet it within minutes. One of your best defenders is the *macrophage* (which translates as "big eater"), a swollen blob of a cell that sucks in almost any foreign particle that crosses its path, including dead cells, debris, and pathogens. Once enveloped, a battery of powerful chemicals attacks the foreign particle, destroying it in minutes. A typical macrophage may swallow some hundred bacteria before it dies, finally done in by its own toxic chemicals.

> **Note** A macrophage is a type of *white blood cell*. All white blood cells are immune system soldiers—they simply use different tactics. (You'll learn about other immune system warriors, including lymphocytes and natural killer cells, as you read this chapter.) Your body creates all your white blood cells in the marrow of your bones (page 95).

A macrophage

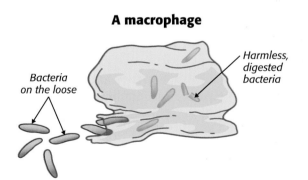

Bacteria on the loose

Harmless, digested bacteria

For its first strike to be successful, your body needs to respond quickly and with overwhelming force. It's not enough to wait until wandering macrophages stumble across new invaders. Your body needs to summon its defensive forces in a hurry.

Its trick is the *inflammation response*, your immune system's call to arms. Your body triggers inflammation when it detects damaged tissue, intense heat, dangerous chemicals, or potential attackers. The first effect of the inflammation response is increased blood flow—your blood vessels dilate and gaps open in your cell walls so the blood can pour into the surrounding tissue. As the blood rushes in, you feel the resulting swelling, as well as pain (because the swollen tissues press on nearby nerves that carry pain signals) and heat (because of the influx of heated blood).

The main goal of the inflammation response is to stock the affected area with your body's immune system soldiers. In addition, the added blood increases the heat, which spurs your macrophages to work harder and can alter the delicate balance of chemical reactions in the invading pathogens, throwing them off balance. (Your body uses a *fever*—a sudden spike in body temperature—with much the same effect when battling more stubborn enemies.)

Note The only time you see your white blood cells is when *pus* oozes from a wound. This creamy, yellow substance contains the detritus of biological warfare—broken-down tissue cells, living and dead pathogens, and scores of dead white blood cells.

The Third Front: Tactical Forces

Inflammation is a brute-force response that overwhelms ordinary microbes. But some pathogens are clever enough to deal with the onslaught of macrophages and the other conventional forces of your immune system. They use a number of subversive tactics to hide, resist engulfment, or prevent your immune system cells from communicating. Some pathogens can battle macrophages from the inside, avoiding digestion or drilling an exit passage. And the meanest microbes launch their own attacks against your immune cells. Your defenses and their countermeasures are all part of an endless biological arms race.

Improving Your Immune System

The most successful ways to help your immune system is through prevention and protection—in other words, practicing tactics like washing your hands properly, avoiding spoiled food, and staying up-to-date on your vaccines. Aside from vaccinations, none of these strategies actually affects your immune system. Instead, they help you fend off pathogens before they get inside.

Although there's no clear-cut way to beef up your body's defenses, some simple practices may help.

- **Eat a balanced diet.** There's no immune system-boosting superfood, but a balanced diet is a good basis for fighting disease. Scientists link deficiencies in folic acid and in vitamins A, B, C, and E, to poor immune-system function.

- **Avoid stress.** And not just the mental kind. The physical wear and tear of weight-loss diets and intense exercise also damps down your immune response. So do injuries, infections, heavy drinking, and pack-a-day smoking.

- **Get a proper night's sleep.** It's obvious, but worth repeating—studies find that 8 hours of nightly sleep boosts the level of immune cells in your body.

- **Be optimistic.** It sounds a bit foofy, but studies have consistently found that you can get someone's body to produce more T cells if you force them to meditate, relax, watch funny TV, or hang out with friends.

To deal with these more resourceful attackers, your body uses another approach, which is called the *adaptive immune system*. It works through specialized white blood cells called *lymphocytes*, so called because they spend much of their time patrolling the passages of your lymph system (page 168).

Your body has two types of lymphocytes:

- **B cells.** When these cells discover an attacker, they produce huge amounts of *antibodies*—small molecules custom-designed to neutralize one specific pathogen and nothing else. All antibodies work by latching on to their target pathogens, but they have different effects. Some disarm microbes immediately. Others act as markers—they flag invaders so other parts of the immune system can find and destroy them.

- **T cells.** These cells form the crack squadron of your immune system. Their specialty is tracking down and destroying cells that harbor viruses. (This is important because antibodies can't penetrate infected cells.) There are several flavors of T cells, including some that shut down your immune response once the attacker is destroyed.

The second and third lines of defense in your immune system complement each other. The front-line forces are always ready to react to a new threat, and they keep many dangerous invaders temporarily in check. This buys time for the adaptive immune system, which takes several days to get up to speed. But once your immune system primes its B cells and T cells, it can mount a perfectly tailored, devastating counterattack.

Acquired Immunity

Without a doubt, the best feature of your immune system is its staggering memory. After defeating an enemy, your body keeps a few specialized B cells and T cells called *memory cells*. These cells remember the identity of their vanquished foe. If that foe returns, even years into the future, your body uses these memory cells to quickly create new copies of the germ-fighting B cells and T cells to attack this repeat offender.

In fact, your body's response to a repeat infection is usually far stronger and more decisive than the initial one. The ability of your immune system to train itself to destroy the enemies it's met before is called *acquired immunity*.

> **Note** Acquired immunity won't help if you meet a subtly different version of the same pathogen. Some pathogens mutate so quickly that they always present a new challenge to your immune system. The most obvious examples are the stubborn viruses that cause influenza and the common cold.

Your immune system has the astounding ability to remember millions of enemies. However, the downside is obvious—to acquire the right memory cells, you first need to get the disease. One way to get all of the immunity with little (or none) of the sickness is to acquire a milder version of the same disease.

For example, in the 18th century, milkmaids who contracted a relatively mild case of cowpox found themselves immune to the much deadlier smallpox. Vaccinations are a modern take on this principle—they prime your immune system with pathogens that are dead or have been rendered chemically harmless. This is enough to educate your immune system without making you sick. (Although sometimes your immune system will get a little overexcited as it reacts to the foreign substances, causing a mild fever, rash, or inflammation.) Thanks to vaccines, smallpox is extinct on our planet, so there's no excuse for cavorting with the milkmaids or their cows.

Another way to pick up immunity without any suffering is to get the pathogen-fighting antibodies. Certain temporary vaccines work this way—they're filled with antibodies that your body can use directly, but temporarily. In the natural world, the ultimate example of this is breast milk, which passes a soup of antibodies from mother to child. Once again, the immunity is temporary. In time, the antibodies break down, and infectious agents will challenge the child to create a personalized set of B cells and T cells, thereby ensuring lifelong immunity.

Frequently Asked Question

Is There a Link Between Vaccines and Autism?

It's a nightmare scenario for concerned parents—a routine vaccination triggers a lifelong developmental disorder that can impair communication and social function. But is it a genuine risk, or is it just the wild paranoia of conspiracy theorists with tinfoil hats?

As any self-respecting scientist will tell you, absolutely anything is *possible*. What seems stark-raving bonkers today just might become the cornerstone of new discoveries tomorrow. That said, scientists have spent a solid decade searching for a link between vaccines and autism, and to date, a great deal of peer-reviewed research has conclusively found no relationship. The removal of a much-feared mercury additive from vaccines has had no effect on rates of autism. And promising new lines of research suggest that the roots of autism may be set by birth, and possibly influenced by events in the womb.

Today most scientists attribute the rise in autism rates to increased reporting—in other words, now that the disorder has a name and a clear identity, parents and doctors are quicker to spot it. One thing is clear: While avoiding vaccines is unlikely to protect against autism, it does roll the dice on an unsettling collection of childhood illnesses. Many of these diseases can kill or cause debilitating effects that last a lifetime—a fact that has slipped out of modern consciousness as the last people who faced those diseases die of old age.

Furthermore, it's important to realize that parents who refuse to vaccinate their children have it relatively easy today, because the most likely avenues of infection—other people—are themselves mostly immunized. But there comes a critical point where a community has enough non-immunized people to sustain a deadly infection and pass it around. In Britain, where the immunization rate has recently dipped to 85 percent, a young boy who had not been given the measles-mumps-rubella vaccine achieved a dubious distinction: He became the first British citizen to die of measles in 14 years.

Bacteria

Feeling lonely? By the time you finish this section, you'll have a somewhat different perspective. Even if you've found a nice, secluded spot to curl up with this book, you're never far from trillions of life-forms—in fact, your body is throwing them the best party around.

Your most successful cohabitants are *bacteria*—microscopic organisms that have only a single cell to their names. Their numbers are overwhelming—you'll find millions in a drop of saliva and billions in a gram of garden soil. And despite the fact that you'll never see one (at least not without help from a seriously powerful microscope), they're easily the world's most populous life-form. As microbiologists like to say, *we* are the minority on planet Earth.

Up close, bacteria have a variety of shapes. They may look like tiny spheres, spirals, or—most commonly—stubby rods, like the happy family of *E. coli* shown here:

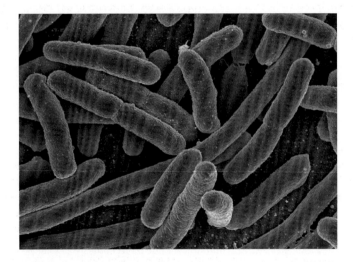

Bacteria of all kinds cover just about every livable surface in the world around you. In fact, they thrive in a lot of places you probably don't want them, such as your kitchen silverware. But the part that you really don't want to think about is that you're never far from their favorite home—*you.*

Bacteria colonize your mouth, throat, and eyes. Entire civilizations of exotic life-forms reside on your skin. But most of the bacteria in your body live in the winding passages of your digestive tract. In fact, bacteriologists say that the number of bacteria in your intestines is *10 times* greater than the total number of cells that make up your entire body. (Fortunately, bacterial cells are quite a bit smaller than the other cells in your body.) This fact has led some sharp-witted scientists to ask if our bodies are really designed for us, or if we're just around to serve as a giant luxury hotel for the care and feeding of bacteria.

> **Note** Next time you look at your weight on the scale and see a number that's a shade too high for your liking, remind yourself that bacteria account for a solid 4 pounds of the total.

Bacteria are responsible for many infamous diseases, including anthrax, tetanus, tuberculosis, the Black Plague, and (on a more embarrassing note) syphilis and gonorrhea. Bacteria are also behind most food-borne diseases, including one of the world's most studied pathogens, *E. coli*.

Late-Night Deep Thoughts

You *Are* Bacteria

We're used to thinking of bacteria as unwelcome hitchhikers. But in many cases, the boundary between the human body and bacteria is surprisingly blurred.

First, your microbial partners have a hand in many of the functions that we consider a normal part of human life. Without beneficial bacteria, you'd lose valuable vitamins (page 196) and give up your unique, lifelong body odor (page 27). You'd also have no defense against the attack of far more dangerous strains of bacteria and, as a result, you'd probably suffer from nearly constant diarrhea, skin rashes, and bladder infections. It's even possible that without friendly bacteria to continuously prod your immune system, its careful calibration would slip, leaving you at a greater risk for allergies and asthma (page 154).

But bacteria are more than the permanent residents of your body. There's solid scientific evidence that the relationship between humankind and bacteria is older and far more complex than most people realize. Many scientists now believe that *mitochondria*, the biological power plants that fuel the work of every human cell, were once free-floating bacteria that somehow became incorporated into the bodies of our most ancient, primitive ancestors. And almost all scientists agree that the tree of life leads back through time to a simple, single-celled creature that was at least a bit like a modern-day bacteria, which means that the stomach bug that makes you ill today might bear more than a passing resemblance to your great-great-great-great- (and so on) grandfather. Thank goodness no one had invented antibacterial soap back then.

Skin Bacteria

If you don't want to come into contact with bacteria, your first step is to avoid touching yourself. That's because healthy human skin is teeming with dozens of different types of bacteria that are all competing for a plot of prime skin real estate. Some help you, some hurt you, but most just hitch a harmless ride that lasts your entire life.

So what do we know about the bacteria that call our bodies home? Surprisingly little—in fact, scientists spend most of their time concentrating on species of bacteria that cause disease. But if you're really intent on making yourself unpopular at parties, here are a few things you should know about your bacterial cohabitants:

- **They're unique.** Each of us has a different blend of bacterial species living on our skin. In fact, studies show that only about 13 percent of the bacterial species on your left hand are shared with your right hand. Based on this principle, some forward thinkers imagine a day where microbiologists can examine objects and determine who touched what by sampling the bacteria they've left behind.

- **They have preferences.** Specialized bacteria live in different ecosystems on your body. For example, your inner elbow has its own thriving bacterial communities that are quite different from those on your inner forearm, only a few inches away.

- **Women's hands have more germs.** No one's sure why women carry around more hand bacteria, despite the fact that they wash their hands more frequently. Possible explanations are the lower acidity of female skin or differences in sweat and skin oil.

- **Washing won't remove your permanent settlers.** It may cut down their numbers (particularly if your soap includes an antibacterial compound that stays on the skin), but they'll soon reestablish themselves in their normal proportions. In fact, bacteria may spread more aggressively immediately after a deep cleaning—which is good, because you need some bacteria to keep your skin healthy, as described in the next point.

- **Competition is good.** The bacteria on your body protect you from more dangerous, disease-causing strains. That's because the benign bacteria that's packed onto your skin doesn't leave much room for anyone else to get established.

Living in a Bacterial World

Invisible bacteria are everywhere. Trying to find their many hiding places is like counting poppy seeds in a bagel factory. And while having bacteria on your skin is a healthy fact of life, the bacteria that fill the world around you aren't always as well behaved.

In a sense, we have no one to blame but ourselves. Bacteria mostly stay put, until a person like you arrives to move them around. All day long, while we think we're cooking, cleaning, looking after our kids, or working at the office, we're also busy transferring bacteria from one place to another. To find the favorite living spaces for bacteria, you simply need to look at the places we touch most.

For example, in a public restroom the amount of bacteria on the much-feared toilet seat is minimal. It certainly can't compare to the thriving colonies on the sink taps and door handles. And in many popular restaurants, the bacteria in the ice machine top what you can extract from toilet-bowl water—a fact originally discovered in a 7th-grader's science project. (The toilet has the advantage of frequent cleaning and fast-running fresh water to rinse it out. The ice machine has the disadvantage of coming into contact with countless people's grubby fingers and a much less frequent cleaning schedule.)

Tip Before you panic, remind yourself that bacteria can't cause any trouble until it breaches your body's defenses. So that means it's probably safe to touch virtually anything in the filthiest restroom, as long as you give your hands a thorough washing before you poke a finger in your eye, mouth, nose, or bacon sandwich.

Your own home has similar surprises in store. The places you come into contact with when you touch and prepare food—such as kitchen sponges and rags, cutting boards, and countertops—as well as doorknobs and toothbrushes, are the most bacteria-laden. In fact, if an alien being were to arrive in your home, it would have to be excused for using the kitchen sink as a washroom and preparing a cheese plate on the toilet seat. From a microbiologist's point of view, this arrangement would be safer for everybody.

Note One study found that bachelors have the cleanest kitchens. That's because they're less likely to use the kitchen to prepare food, and even less likely to pick up a rag to clean off a countertop (which often simply smears the bacteria around).

So with all this bacteria on the loose, what's a paranoid person to do?

- **Sanitize sponges and cutting boards.** These are not only hotspots of bacterial life, they're also the primary vehicles for spreading bacteria around. To sterilize a sponge, you can boil it in hot water for a few minutes, run it through a dishwasher drying cycle, or soak it lightly and then pop it in the microwave. To clean your cutting board, scrub it with a light bleach solution (1 tablespoon of bleach to a quart of water). To be extra safe, use a separate cutting board for meat duties, and replace your cutting board when it becomes heavily scored, because bacteria love to pile into the grooves.

- **Treat the sink with caution.** Give it a thorough, regular cleaning. And once you clean the sink, make sure you dry it thoroughly. That's because a moist environment encourages the last, lingering bacteria to reproduce. And whatever you do, don't eat something you've dropped into your sink unless you cook it first.

- **Prepare food properly.** Proper cleaning is important for foods that you plan to eat raw, like fruits and veggies. To prevent cross-contamination, rinse all fruits and vegetables, even those that you plan to peel (like oranges, cantaloupes, and potatoes). Rinse with ordinary water—soap may leave a residue you shouldn't ingest, and fancy food-cleaning systems don't make much of a difference in serious tests. Lastly, don't wash chicken or raw meat before you cook it. Your oven destroys all the bacteria it contains. Washing the meat simply gives you an opportunity to spread tiny droplets of bacteria-laden water throughout your kitchen.

- **Wash your hands properly.** With a world of bacteria around you, it's safe to assume that your hands harbor some unwanted guests. Clean your hands before you handle food and before you sit down to eat. Soap and warm water does the trick. Scalding hot water still won't be hot enough to kill bacteria, so don't torture yourself. Antibacterial soap isn't much help, either. The active germ-killing ingredient, *triclosan*, works only if you leave the soap on your hands for several minutes before rinsing, which virtually no one does.

> **Note** Studies show that ordinary soap does a perfectly good job of removing dangerous pathogens from your hands. Antibacterial soap can do the same job, but it has a potential side effect. By rinsing antibacterial chemicals down the drain, you increase the odds that they'll encounter a colony of bacteria on the way down and cause it to evolve into a more dangerous, resistant strain. If you use antibacterial products, choose the ones that actually have proven benefits—for example, toothpaste (page 184). And skip the antibacterial soap, which offers nothing more than a dose of false comfort.

Is Raw Cookie Dough a Killer?

For many, baking cookies is a labor of love that's sweetened by the occasional stealthy scoop of raw cookie dough. But there's a sinister side to this guilty pleasure: Public health officials warn that raw eggs can contain stomach-churning *salmonella* bacteria, which can cause fever, diarrhea, and even death. So should cookie bakers keep their fingers to themselves?

First, it's important to realize that no one really knows how many eggs are contaminated with salmonella. In the past, experts thought that salmonella lived on eggshells, but couldn't make its way into an egg without traveling through a hairline crack. Today we know that salmonella can pass from the ovaries of infected hens straight into their developing eggs.

In the Northeastern states of the U.S., solid estimates suggest that 1 in 10,000 eggs are contaminated with salmonella. That means you could eat an entire batch of two-egg cookie dough and face only a 0.02 percent chance of a night on the toilet. Of course, these figures are only estimates—some studies put the number of infected eggs at 1 in 20,000, while at least one ratchets it up to 1 in 700.

Even then, tainted dough may not be as dangerous as it seems. Studies show that salmonella needs the power of numbers to wreak its damage, and healthy volunteers don't get a serious infection unless they ingest about 1 million salmonella organisms. (This is notably different from dangerous strains of *E. coli*, which can breach your body's defenses in very small numbers—as few as 200 bacteria.) And if you do get infected with salmonella, the odds are overwhelming that you'll be back on your feet in a week with nothing worse than some painful memories.

The bottom line? Eating raw cookie dough is particularly risky for young children, pregnant women, the elderly, and people with impaired immune systems—all of whom are more likely to suffer dangerous complications. (And to be consistently paranoid about egg safety, none of these individuals should eat a runny-yoked egg, which may still harbor bacteria.) But an average, healthy adult with a normally functioning immune system has a relatively small risk of serious health trouble. On the other hand, exercise caution when dealing with foods that traditionally use raw eggs—such as Caesar salad dressing, eggnog, and homemade ice cream. These foods aren't eaten immediately, which gives bacteria time to multiply and reach more dangerous levels. To keep these foods safe, make them with pasteurized egg products.

Antibiotics

The standard way to combat a bacterial infection is with *antibiotics*, a class of chemicals that kills bacteria. Although we group them into a single category, different antibiotics work in different ways. Some destroy the bacteria's cell walls, causing them to burst and die. Others disrupt the processes bacterial colonies use to reproduce. Either way, the basic principle is the same—antibiotic drugs interfere with the machinery of bacterial life without affecting the way human cells operate.

> **Note** Just as antibiotics have no effect on human cells, they also leave viruses, parasites, and fungal infections untouched. You'll need different types of drugs to battle these attackers—antibiotics will have no effect. And in the case of viruses, you'll usually be forced to wait and suffer until your immune system ramps up its defenses. (That's why a trip to the doctor's office won't help you cure the average cold or flu.)

Some antibiotics work against certain families of bacteria and are called *narrow-spectrum antibiotics*. Others destroy wide swaths of bacterial life and are called *broad-spectrum antibiotics*. But neither sort can distinguish between the bacteria that harm your body and those you'd like to keep. When antibiotics wipe out the beneficial bacteria in your colon and on your skin, they often lead to side effects like diarrhea and fungal infections of the mouth, digestive tract, and vagina.

A more serious problem is *antibiotic resistance*—the ability of bacteria to evolve immunity to commonly prescribed antibiotics. Antibiotic resistance usually occurs when a colony of bacteria meets up with antibiotic drugs. Although these antibiotics destroy virtually all the bacteria—and they do it quickly—they may leave behind a few rare mutants that have some level of natural immunity. If the antibiotic attack keeps up, these mutants will eventually die along with their weaker relatives. But if the onslaught ends, these mutants will have a chance to establish a new, more resistant colony. Repeat the process a few times, and you'll gradually breed stronger and more resistant bacteria. And throw in a few different types of antibiotics, and you just might produce a superbug that's impervious to all forms of standard treatment. Even worse, bacteria have a naughty habit of swapping DNA, which means the antibiotic resistance that develops in one species can leap to another, more virulent strain.

Now that you understand antibiotic resistance, you know why your pharmacist always tells you to finish the full course of your antibiotic prescription. If you have an infection, most antibiotics will destroy the large majority of bacteria in just a couple of days. But if you stop at that point, you may spare a few, resilient stragglers. And like the son who avenges the father in a cult karate movie, those bacterial stragglers just might come back to wreak some serious havoc in your body or someone else's.

Note Another way to help prevent antibiotic resistance is to avoid using products that contain unnecessary antibacterial chemicals. The best example is antibacterial soap, a mostly useless product that's debunked on page 216.

Frequently Asked Question

Can You Taste Spoiled Foods?

E. coli and other food-borne bacteria (like salmonella and campylobacter) are tasteless. Contaminated food gives no obvious sign of the single-celled organisms that lurk there, waiting to colonize your body. This raises an excellent question: If dangerous bacteria don't leave obvious signs, what's making that week-old package of ground beef smell so bad?

The answer is *spoilage bacteria*, a family of bacteria that thrives on just about any food. As these bacteria replicate, they coat your food with slime. The waste products they leave behind cause the objectionable changes in smell and taste. However, for all their obvious repulsiveness, eating them probably won't make you sick. That's because spoilage bacteria is ideally suited to the world of decaying grocery produce, not the high-acid environment of your stomach.

So does this mean that you can add rotten food to your dinner table without harm? Well, not quite. As spoilage bacteria break down food, other organisms hitch a ride. Molds quickly join the party (for example, the fuzzy green fur on forgotten salami). Some are dangerous and create poisonous substances that permeate food, and they remain even after cooking.

Furthermore, if conditions are ideal for spoilage bacteria, it's a safe bet that they're good for pathogenic bacteria, too. In other words, if your food is spoiled, it's also more likely to hold a teeming population of pathogenic bacteria, and therefore to pose a greater health risk.

Viruses

In some respects, the battle between bacteria and your body is refreshingly straightforward. Huge armies of microscopic, foreign creatures flood your body. They wreak havoc—briefly—before your better-armed defensive forces destroy them.

Viruses are a different matter. First, they're much tinier—about a hundredth the size of an average bacterium. In fact, viruses are so vanishingly small that even the most powerful optical microscope can't spot one (although a cutting-edge *electron microscope* can). Stack viruses and bacteria together, and it's like comparing a toddler to a brontosaurus.

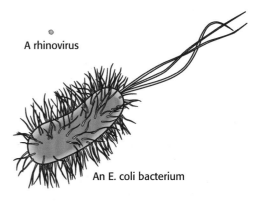

A rhinovirus

An E. coli bacterium

> **Tip** To get a better feel for the difference in scale, check out *www.cellsalive.com/howbig. htm*. There you'll see a simple animation that places you on the head of a pin and increases the magnification until you can spot a dust mite (page 11), a particle of pollen (page 150), and a red blood cell (page 136). Zoom in still more and you'll be able to make out the much smaller cell of a bacterium, and then, finally, a virus.

There's another clear difference between bacteria and viruses, but you need to step into their microscopic world to see it. Up close, a bacterium looks like a tiny alien being. It may be small (and ugly), but it's full of life—feeding, reproducing, and generating energy with some of the same processes your own cells use. Many bacteria are even able to move by propelling themselves with long, whip-like tails, or by gliding along paths of self-produced slime.

By comparison, a virus looks more like a piece of organic debris. Its structure is simple—in fact, a virus consists of little more than a submicroscopic scrap of genetic material (either DNA or its relative, RNA) wrapped in a thin coat of protective protein. On its own, a virus is silent, inert, and completely lifeless. It's unable to power a single one of the chemical reactions required for life.

If it weren't for the presence of other life-forms, this is where the story would end. But as you'll see, viruses have the uncanny ability to turn up at the right place at the right time—namely, in the midst of a normal cell's manufacturing process.

The Life Cycle of a Virus

To do anything, a virus needs your help. First, it needs to get into your body, and it does that in much the same way as bacteria does—by being inhaled into your lungs, swallowed into your digestive tract, or absorbed through a cut in your skin.

Once inside your body, the virus drifts aimlessly until it comes into contact with the right cell—one that has a coat of proteins that complements those of the virus. When the virus bumps up against this cell, its proteins lock on. (Keep in mind this isn't a conscious decision for the virus—it's simply a reaction caused by the fact that it fits the target cell like a fuzzy sweater and a strip of Velcro.)

What happens next is more unsettling. The virus launches the multiple steps of its attack procedure, shown in the following figure:

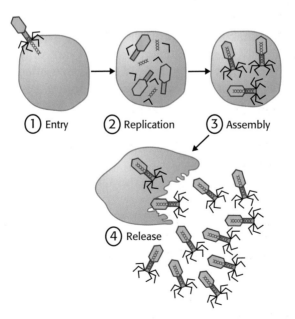

(1) Entry   (2) Replication   (3) Assembly

(4) Release

First, the virus needs to get inside the cell. In some cases, the cell may engulf the virus in the same way that it swallows tiny nutrients, pulling it in. Or the virus may inject its genetic material through the cell wall. Either way, the damage is done. The foreign genetic material finds its way deep into the cell's working parts, where it quickly takes over.

In the classic case, the virus inserts a short snippet of DNA into the target cell (as shown in step 1 of the figure on page 221). Like all strands of DNA, this DNA contains instructions for building specialized proteins. The target cell cheerily follows these instructions, unaware that it's helping the enemy.

Once built, these proteins begin to carry out their pre-programmed functions—manufacturing thousands of new viruses. (This is what happens in step 2 and step 3.) In this way, the virus hijacks the inner workings of the cell, like a pirate commandeering an ocean liner. But all the while, the virus hasn't actively done a single thing. It's just a set of malicious instructions that your body executes, simply because the virus was in the right place at the right time.

Once they've taken over a cell, most viruses replicate like Viagra-fuelled rabbits. Eventually, they leak out of the cell through tiny pores or blow it apart like an overfilled water balloon (as you can see in step 4).

There's a virus for virtually every type of cell. Viruses infect animals, plants, and even bacteria. (In fact, it's likely that the bacteria that causes cholera would be completely harmless were it not for the presence of a toxin-producing virus embedded inside.) However, the viruses that affect one species are often unable to affect another, or they may have dramatically different effects. HIV is a well-known example—not only is it unable to infect other animals, but the related SIV strains that affect monkeys and chimpanzees rarely cause the compromised immune system and debilitating symptoms of AIDS.

Note Viruses don't necessarily correspond to illnesses. In fact, many viruses have no symptoms. They don't destroy their host cells, reproduce very quickly, or create poisonous compounds. Virtually all people have at least a few harmless viral passengers hiding in their bodies.

Famous viruses include smallpox, rabies, polio, measles, and the common cold, which you'll consider next.

You *Are* a Virus

Not long ago (see the box on page 213), you learned that many of the basic processes of human life require a partnership between your body and the bacteria that calls it home. Now you're ready to learn another disturbing truth: you have an even more intimate relationship with viruses.

The key players are *retroviruses*, a broad class of viruses that carries strands of RNA instead of DNA. The interesting and highly technical part is that your body sometimes converts RNA back into DNA. As a result, a virus that holds a piece of RNA is able to fuse itself into your genes, altering your genetic code.

Before you panic, remember that your body has trillions of cells, and virus-infected cells generally don't last long before white blood cells destroy them. However, there's always the possibility that a virus will find its way into a *germ-line cell*—in humans, these are the cells that produce male sperm and female egg. If a virus lands in one of these cells, it has a good shot at being incorporated into the genes of the next generation. And that's not science fiction, as recent studies suggest that nearly 10 percent of human DNA consists of pasted-in viruses from the past.

Now, it's important to understand that a virus integrated this way probably isn't going to *infect* newly minted babies. That's because the virus incorporates its DNA in a random place. Usually, it's in the vast wasteland called *junk DNA*—segments of your genetic code that don't appear to do anything at all. However, occasionally a virus lands somewhere important, and the result is usually trouble. Some researchers believe hemophilia and muscular dystrophy are two genetic diseases that cropped up when random, viral DNA blundered into the wrong spot.

Finally, here's the really interesting part: As you probably know, evolution works when apparently random changes in a creature's DNA give it a valuable survival advantage. And while viral DNA is more likely to cause a problem than to confer a benefit (and is most likely to do nothing at all), every once in a while a bit gets into a place where it just might do real good. In fact, many scientists believe that viruses have helped shape evolution on our planet by reshuffling genes, pasting in their own contributions, and carrying genes from one species to another. So while life started as mere bacteria, viruses just might have supplied some of the variety that drove evolution forward and led, eventually, to the creation of you.

Profile of a Cold

One type of virus that your body knows intimately is the one that causes upper respiratory tract infections, which are otherwise known as the *common cold*.

Colds appear to expose a chink in the defenses of your immune system. After all, the average person suffers three or four colds a year, and no matter how many you endure, you're never rewarded with lasting immunity. The reason for this endless suffering is variety. Scientists recognize more than 200 viruses that cause colds, and it's likely that there are many more on the loose, unknown and uncataloged.

This diversity raises an obvious question: How can so many different viruses cause essentially the same symptoms when they infect you? The answer is that the symptoms of a cold aren't caused by the virus itself, but by the inflammatory response that your body greets it with. This often starts with pain and swelling in the throat, followed by a runny nose as your body attempts to wash out virus particles. If the inflammation makes its way deeper into your throat, the next inflammatory symptom is coughing.

> **Note** The viruses that cause a cold are successful for several reasons. First, they mutate quickly and effectively, challenging your body with an endless assortment of slightly different variations. Just as important, they're easily transmitted and unlikely to kill you. This last point is critical, because viruses have the best chance to get around if they keep you alive and healthy enough to walk around spreading them to everyone you meet. In fact, many of the viruses that we fear most—dramatic, suddenly lethal ones like the Ebola virus—are actually profound failures. They kill infected people so quickly that the viruses become trapped in dying bodies.

The odds are that you'll spend some time this year battling at least one cold. Here are a few tips to keep in mind:

- **Colds aren't an indication of poor health.** We all know someone who makes it through the year without the faintest sniffle—and someone else who spends an entire month bleary-eyed and runny-nosed. The odd truth is that both people may be catching the same cold viruses, but simply experiencing them differently. Before you envy the person who slips by with nary a symptom, remember that a laid-back immune response can allow a cold virus to spread farther and even cause damage before it's destroyed.

- **Vitamin C doesn't help.** It's an enduring myth, but countless studies show that there's basically no benefit to the citrus vitamin. The exception is marathon runners and people who perform strenuous exercise in the cold, where vitamin C appears to reduce the risk of catching the cold virus (but still does nothing to cure an existing cold).

- **Blowing your nose can be risky.** Most scientists agree that blowing your nose doesn't provide any benefit for your body (other than comfort). However, there's a darker side to nose blowing. As you learned when you explored nasal mucus (page 144), overly vigorous nose blowing can drive viruses and inflammatory substances into your sinuses, possibly causing additional pain or infection.

- **Colds travel through snot.** You most commonly pick up the cold virus through airborne droplets of mucus (generated by someone else's sneeze), or by touching a contaminated surface. Kids are prime transmitters, but even adults are adept at transmitting nearly invisible traces of mucus from their noses to their hands and then to everything else in the surrounding environment. However, the cold virus still needs to jump through a weak point in your body armor, such as your eyes, nose, or mouth. So after you touch any of these vulnerable places, make sure you wash your hands.

- **Colds might prime your immune system.** There's no cure for the common cold on the horizon. Even if there were, you might not want to take it. Some researchers believe that a cold-free life might leave people at increased risk for allergies and asthma. (To learn why, see page 232.)

Frequently Asked Question

What Is the Deadliest Virus?

When it comes to deadly viruses, Ebola kills in the quickest and most horrific way possible—causing massive bleeding and turning internal organs into a soup of lifeless mush. It's estimated that 90 percent of Ebola-infected people die from these symptoms. However, HIV (the virus that causes AIDS) is still more effective—eventually, virtually everyone who contracts HIV will have their immune systems knocked offline, as the virus infects the very T cells that are supposed to defend the body. Without treatment, a person suffering from AIDS is unlikely to last even a few years, as hundreds of ordinarily harmless microbes ravage the body.

However, neither of these viruses can boast the highest body count through history. That dubious distinction probably belongs to influenza, the virus that causes the flu. Each year, influenza kills hundreds of thousands of people across the globe, most often the very old or the very young. But every few generations, a strain appears that is far deadlier, like the 1918 Spanish flu, which killed tens of millions of people in a single, worldwide outbreak.

Cancer

So far, you've spent most of your time exploring battles that are relatively clear-cut. They pit your body against outside forces, like invading bacteria and viruses. But now it's time to consider a subtler enemy—rogue cells in your own body.

The disease is *cancer*, and it starts with a subtle genetic shift that transforms one of the trillions of ordinary cells in your body into a saboteur. Unlike the cells in the rest of your body, a cancerous cell isn't bound by the normal rules of human life. It grows without respect for the boundaries between different types of tissue, invades other sites in the body, and refuses to die a natural death. In the end, what begins as a series of simple cellular errors can become an unstoppable process that ravages your body.

How Cancer Starts

Although we often imagine cancer as a single thing, it's actually a family of diseases that's characterized by misbehaving cells. The problem begins with a chance mutation in a key regulatory gene—essentially, a cell turns off one of the safeguards that restricts it or over-activates one of the mechanism that drives normal cell growth. However, a single mutation isn't enough—if it were, you'd be riddled with cancer while you were still in diapers. Instead, cancer needs to develop through a succession of highly improbable mutations, which gradually give the cell and its offspring the ability to defeat several different control mechanisms.

The picture on the next page shows one way this process can unfold:

1. The cell begins its life as normal.

2. Random mutations give the cell the ability to ignore the normal recycling processes of your body, so instead of dying, the cell lives forever. This transformation happens quietly and without event.

3. Next, the cell multiplies, creating similar ill-tempered progeny and crowding healthy cells out of the way. This unchecked growth often creates an abnormal mass of tissue called a *tumor*, which can cause problems if it presses on one of your vital organs.

> **Note** Ordinarily, cells have a built-in self-destruct sequence. When a cell detects that it's diseased or damaged (or when other cells detect something suspicious and convince the cell that it's not quite right), the cell initiates this self-destruct sequence and destroys itself in a calm and orderly fashion. This tidy suicide process is called *apoptosis*, and it's as fundamental to the functioning of your body as cell division. However, successful cancer cells don't obey the shutdown command—they stay alive, multiply, and can develop more dangerous mutations.

4. The real trouble with cancer occurs when the cancerous cells *meta-stasize*, or spread to other areas of your body. Once cancer cells have become mobile, they travel far and wide, voyaging through your blood and lymph (page 168), and starting new cancer settlements throughout your body. At this point, the odds of successful treatment dwindle quickly.

Because different cancers acquire different mutations, they vary in their virulence. The nastiest forms multiply quickly and travel aggressively, and they can rapidly colonize your body. Other forms are highly treatable and have better survival rates than a heart attack or stroke.

When diagnosing a new cancer in a patient, doctors classify how far it's advanced by *stage*. The exact definition of the various stages (and the prognosis of a cancer patient) depends on the type of cancer and its location. But in general, stage I cancers have not yet spread and are usually treatable. Stage II cancers have had some time to develop but have not yet traveled the body, while stage III cancers have made it to nearby lymph nodes. Stage IV cancers are the worst—they've spread to organs throughout the body and are usually untreatable.

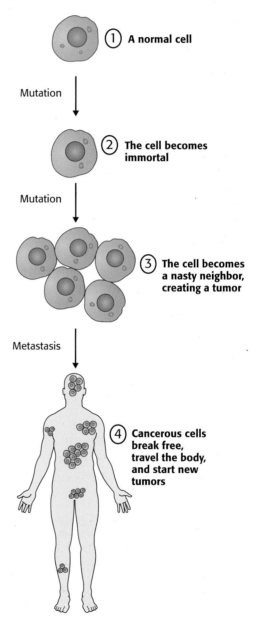

Mutation

① **A normal cell**

② **The cell becomes immortal**

③ **The cell becomes a nasty neighbor, creating a tumor**

Metastasis

④ **Cancerous cells break free, travel the body, and start new tumors**

The Body's Cancer Defenses

Out of all the diseases that affect human beings, cancer is particularly hard to grasp. After all, bacteria and viruses had millions of years to evolve into deadly attackers. Cancer cells are just the product of a random cell gone haywire. So how can they possibly account for the second most likely cause of death in the industrialized world (page 269)?

The answer is bad luck and big numbers. Although cancer-causing mutations are exceedingly rare (on an individual-cell basis), your body has trillions of cells, and it manufacturers millions more every minute. Even a seemingly miniscule rate of cancerous mutation—say, one in a million cell divisions—would guarantee you a terminal case of cancer before you put this book down.

The real question isn't why we get cancer, but why we don't get it more often. In fact, the body has countless built-in safety measures to defend against cancerous cells. It has specialized genes that detect suspicious behavior and shut down cell growth or trigger cell destruction. Your immune system even has a class of specialized warriors called *natural killer cells* that hunt down cancerous cells and release toxic granules that destroy them.

However, none of these mechanisms is completely foolproof, and given enough time (and enough cell mutations), a cancerous cell can start to thrive. Cancer is a particular problem in old age because the body's cells have had more time to accumulate the right mutations, the natural cancer-fighting processes of the body have weakened, and a lifetime of exposure to carcinogens (chemicals in the environment that can damage DNA and spur mutations) has taken its toll. The following chart shows the rate of colon cancer as a function of age, and it tells a clear story—cancer risks skyrocket as time wears on.

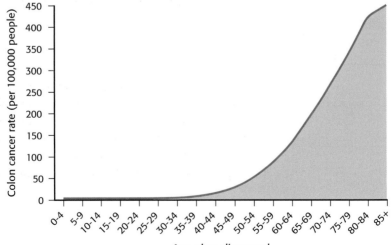

Age when diagnosed

It's all a bit ironic. In one context, sheer mind-boggling odds can turn one mutation in a trillion into the survival advantage that drives the evolution of a species. But in another, a chance combination of deadly mutations can trigger a cancer and destroy an individual.

Cancer Prevention

Now that you understand how cancer works, it's time to ask how you can stop it. Unfortunately, there's no single measure to prevent cancer. In fact, your body is already using all the cancer-prevention programs we know about.

Some cancers are closely associated with particular lifestyle risks. For example, sun exposure is linked to skin cancer (page 14) and cigarettes are tied to lung cancer (page 156). Minimize these risks, and you're likely to avoid the cancers they cause. But many more cancers aren't so clear-cut. They arise spontaneously and unexpectedly, after a lifetime of bodily wear and tear.

Your best bet is to detect the problem early. If you can catch a cancer before it metastasizes, your odds of conquering it are dramatically better. Unfortunately, many cancers have subtle symptoms that aren't initially troubling. Some symptoms—like weight loss, fevers, swollen lymph nodes, or a feeling of constant tiredness—may indicate cancer, but are usually caused by something less serious, like an infection. And a few cancers (for example, pancreatic cancer) are virtually undetectable in their early stages.

To give yourself the best odds, you need to be eternally vigilant for problem signs. Some examples include unexplained lumps, persistent coughing, and blood in your stool. However, it's best to investigate *any* unexpected change in the way your body works. It's also essential to keep surveying the territory with breast self-exams, testicular self-exams (page 237), a yearly physical, and regular colonoscopies after age 50 (or earlier if your family history warrants it).

Flaws in the System

Every hour of every day, you rely on your immune system to keep potential disease-causing agents in check. And for the most part, the system works. Aside from the occasional runny nose (or bowel), the battle between your body and the pathogens that want to infect it rarely disturbs your daily life. In fact, unless something knocks out your immune system—like late-stage AIDS—you're likely to forget all about your body's perpetual fight for survival.

But your immune system is far from perfect. Millions of years of evolution have given it a ruthless, kill-at-all-costs streak, which can wreak serious damage on the peace-loving parts of your body if they get caught in the crossfire. And sometimes, your immune system can veer into outright paranoia and unleash a crushing force against innocent cells or an otherwise harmless substance. The result of this misdirected aggression can be anything from an annoying allergy to a life-threatening disorder.

Autoimmune Diseases

Your body has a savant-like ability to distinguish its own cells from those of foreign infiltrators. This feat may seem straightforward, but it's actually a minor miracle. For it to work, each immune cell—from the lowly macrophage to the highly trained T cell—must be able to quickly and conclusively recognize the hundreds of different cell types in your body, most of which it's never seen before.

To make the distinction between self and non-self, immune cells look for special marker molecules that stud the surface of every cell in your body. These markers are unique—meaning you share them with no one (except an identical twin). As long as your cells have the right markers, your immune system lets them live their lives undisturbed.

Your body's cell-recognizing system is perfectly splendid when it works, which is virtually all the time. However, life is not so smooth for people who suffer from an *autoimmune disease*—a disorder in which the immune system becomes so overly paranoid that it stops recognizing part of its owner's body. The result is that the immune system attacks these unrecognized cells in much the same way that it would attack a microbe or a transplanted organ from another person.

Although autoimmune diseases vary widely in their effects (based on the types of cells the immune system targets), they tend to share a few grim characteristics. Namely, they're chronic, they may flare up and die down over time, and while doctors can treat their symptoms, the diseases themselves are often incurable and probably unpreventable. There are roughly 100 known autoimmune diseases. Some well-known examples include Crohn's disease and colitis (which attack the digestive tract), Hashimoto's disease (the thyroid gland), multiple sclerosis (the wiring of the central nervous system), lupus (the body's connective tissue), and rheumatoid arthritis (the joints). There are still more diseases where an autoimmune reaction is suspected but not proven—from schizophrenia to skin conditions like vitiligo and psoriasis.

It would be great if the right diet or proper exercise could prevent an auto-immune disease, but the reality isn't so sunny. In fact, many autoimmune diseases have a strong genetic link. Experts believe that people inherit the *potential* for a disease, which lingers undetected until something triggers it, such as stress, pregnancy, or another disease.

> **Tip** If you suffer from an autoimmune disease, your best bet for survival is early detection. If left unchecked, the immune system can destroy your body as effectively as it dispatches the remnants of a week-old cold virus. Proper treatment, which may use medication to dampen the damage, is essential.

Allergies

When your immune system attacks the cells of your body, it's considered an autoimmune disease. When your immune system attacks a normally harmless environmental substance (such as dust, pollen, or peanut butter), it's an *allergy*.

At first glance, allergies appear to be harmless. You may not want your immune cells to waste time battling particles of pet dander, but their misdirected hostility won't cause the same damage as an all-out assault on one of your organs. However, the allergic response isn't always innocent, because it goes hand-in-hand with *inflammation*, the body's attack call. Unchecked inflammation can cause permanent damage—for example, in untreated asthma (page 154). The most severe hypersensitivity, *anaphylaxis*, causes such an extreme overreaction that shock and death can follow in minutes.

You probably already know the basic treatments for allergies. You can treat simple cases—sensitivities that cause sneezing, itching, and rashes—with an occasional dose of over-the-counter antihistamines. These chemicals suppress the action of *histamine* in your body, a compound that helps trigger your immune response. Other treatments include avoiding the allergens that cause all the trouble, using more powerful drugs (like steroids), and *immunotherapy* (a 3- to 5-year program of allergen injections that aims to gradually desensitize the immune system). Immunotherapy doesn't work for everyone, and it's usually reserved for people with extreme, potentially life-threatening sensitivities. It's also the only approach that aims to cure the problem rather than treat the symptoms.

The Age of Allergies

Allergies fascinate modern researchers because they're on the rise, particularly in rich, developed countries. When looking for potential causes, many experts blame a few things we've come to enjoy—our high standards of cleanliness, our vaccines, and our heavy use of antibiotic drugs. In other words, by bringing modern medicine in to help us fight our battles, we've disrupted our long-standing relationship with the microbial kingdom.

Back in the good old days, when the average person lived in squalor, and illness killed nearly a quarter of all infants, things were different. Yes, life was often nasty, brutish, and short—but the onslaught of potentially deadly pathogens gave your immune system a well-rounded education in its formative years. If you made it to adulthood, you probably didn't have to worry about hay fever.

By comparison, the children of today play in relatively sterile environments. They're immunized against serious childhood diseases—from measles to polio—that claimed countless lives only a few decades ago. And if their immature immune systems come up against a stubborn infection, we use antibiotics to tip the odds in their favor, possibly wiping out the beneficial bacteria of the large intestine at the same time. These measures just might disrupt the natural maturation of the immune system, leaving it overly reactive and prone to allergies and asthmas.

If this is true, then the best way to reduce allergies is to stop fighting microbes with medicine and just let nature take its course. Of course, the inevitable consequence is that scores of people will die at the hands of the meanest viruses and bacteria. Natural doesn't always mean good.

> **Note** That's not to say that there aren't some simple measures that might reduce the problem. For example, avoiding overly prescribed antibiotics, limiting the use of unnecessary antibacterial products, and allowing young children to play in "dirty" places and daycare centers may have some beneficial effects. However, no one knows how well these measures might work. As long as we use modern medicine and modern sanitation to prevent pathogens from killing us, we just might have to accept a few unpleasant side effects.

This idea—that modern medicine or modern hygiene is partly to blame for the suddenly skyrocketing rates of allergies and asthma—is still controversial. However, the balance of evidence suggests that some environmental factor in modern, industrialized countries is the cause. Another theory is that the immune system is naturally a bit over-the-top because it's designed to keep working even in the presence of parasitic infection. In fact, some studies suggest that the lack of allergies in developing countries is due to the fact that just about every adult carries around an immune-suppressing tapeworm or two. Remove the tapeworm, and allergies appear. (One dedicated scientist, Koichiro Fujita, went so far as to proudly swallow tapeworm eggs to battle his hay fever. He claims the measure was successful—and that he's lost some extra weight to boot.)

Note If the link between parasites and allergies is proven true, it's bad news for the natural-living crowd. While it's easy to refuse vaccines or eschew antibiotics, it's hard to imagine a similar dedication about holding onto a 10-foot tapeworm.

10 Sex and Reproduction

own below, not far from the place where digestive waste exits your body, are some organs we're even more embarrassed about. This is a bit odd, because we all have these body parts, and most of us like what we do with them. Still, it's easier to convince people to engage in a 20-minute discussion about the finer points of intestinal gas than to get them to talk frankly about their *sexual organs*.

There was a time, not so long ago, when sex was a bit of a mystery. Now we have call-in health shows, how-to sex videos, and the Internet. But despite all the attention we pay to sex, most of us remain unaware of some key facts about how our reproductive hardware works—whether it's our own body parts or the sexual organs of the opposite sex.

In this chapter, you'll have a good scientific excuse to spend some time thinking about reproduction (and we aren't talking fake Picassos). You'll learn everything you need to know about the baby-making parts of both genders, from the fabled G-spot to the oft-feared vasectomy. You'll also find out what really takes place during sex, and how you can give yourself the best shot at copulatory bliss.

Male Sex Organs

Throughout this book, you've looked at the human body as a single entity shared by people old and young, male and female. Here's where the path diverges.

If you have any giggles, get them out now—we're about to take you back to high school health class.

What's on the Outside

Of the two sexes, men have the most obvious sexual equipment. When you look at a naked man (or, for more modest readers, a classical Greek statue at the museum), you just can't miss the way his sexual bits dangle brazenly out in the open.

The star of the show is the *penis*—a tube of spongy tissue that serves as both a urine-disposal system and a baby-making tool. Thanks to its prominence, the penis has commanded some outsized attention through history. Imaginative minds see it symbolized in office towers, tunnels, weaponry, gushing fountains, sausages, and cigars. The sex-obsessed founder of psychotherapy, Sigmund Freud, thought that everyone—even women—craved a penis of their own. And during the witch hunts of the 14th and 16th centuries, women were tried, convicted, and burned for imagined crimes that included causing impotence, stealing penises, and having "knowledge" of the Devil's phallus.

The most controversial part of the penis is the thin layer of *foreskin* that covers the tip. Few things can heat up a conversation as quickly as politics, religion, and *circumcision*—the snipping of the foreskin from the tip of an infant's penis. Nearly a third of the world's men are circumcised, most often for religious reasons. Advocates promote circumcision as a health measure that can prevent rare forms of penile cancer and slow the spread of AIDS. Detractors call it everything from cosmetic surgery for children to child abuse, and warn that it can deaden the fine sensations in the most sensitive part of the male anatomy.

> **Note** The American Association of Pediatrics offers some balanced circumcision advice in a carefully considered policy statement (available at *http://tinyurl.com/4t72t*). Their bottom line? While circumcision may have health benefits, they're difficult to measure and very minor. From a health and cleanliness point of view, there's no reason to recommend circumcision. On the flip side, if circumcision is part of your cultural tradition, it's unlikely to cause pain or lasting harm.

Behind the penis are two *testicles*, which work like tiny, industrious factories. They create a staggering number of sperm, churning out millions of individual swimmers each hour. They also manufacture key sex hormones like *testosterone*, which boosts sex drive and triggers the body changes of puberty. (This is the reason that *castration*, which is the surgical or chemical removal of sex organs, changes more than a person's sex drive. If you castrate a boy before puberty, he'll forever keep a slender build, hairless face, and high-pitched voice—perfect for singing opera.)

Tip The testicles are also the home of the most common cancer in men ages 15 to 34. Doctors recommend that men examine their testicles monthly to check for signs of testicular cancer, which is easily treatable in its early stages. A testicle exam works a bit like a breast exam—you take the testicle between your thumb and fingers and feel (gingerly) for any lumps.

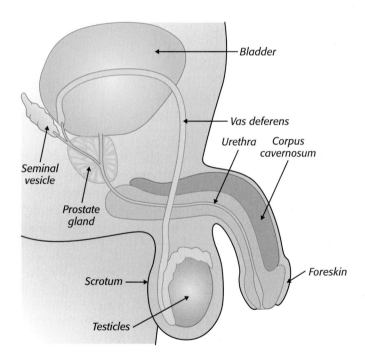

Weak Points of the Male Sex Organs

The male sex organs are a quirky part of the human body—at once practical, impressive, and droll. They aren't without a couple of flaws:

- **Exposure.** Unlike every other gland in the body, testicles dangle out in the open in a sack of skin called the *scrotum*. The key reason is temperature control—by swinging in the breeze, the testicles maintain a temperature that's lower than the rest of the body's, which allows them to produce viable sperm. Unfortunately, this location has a massive downside, which you can spot in the mind-numbing TV show, *America's Funniest Home Videos.* Being out in the open, the very sensitive testicles have no protection against stray tennis balls, flying squirrels, or oversized plastic baseball bats.

- **The prostate gland.** This small gland produces some of the fluid that goes into semen (page 242). As men age, their prostate gland grows bigger. And as the prostate gland enlarges, it may compress the *urethra* (the urine-carrying tube that passes through the prostate gland), making it difficult to pee. Thanks to this disastrous design, half of all 50-year-old men have urinary trouble—and the odds only get worse with age.

Erections

The penis is capable of a fairly impressive parlor trick—when the owner is excited (or at some otherwise unexplainable moment), it nearly doubles in size.

To pull this off, the penis fills two tube-like structures, called the *corpus cavernosa* (see the picture on page 237), with blood. This blood swells the penis in much the same way that air inflates a bicycle tire. Once the penis is fully erect, the arteries clamp shut, trapping the blood inside. Erections involve no muscle contractions, and holding one doesn't take conscious effort. Instead, fluid pressure keeps it firm.

This feat is all the more impressive when you consider that humans lack something that most other animals rely on to stiffen up—a penis bone. The advantage of a penis bone is that erections are lightning fast—the animal in question simply uses a few muscles to slide the bone out of its abdomen and into its penis. In fact, it's the perfect design for animals that need to mate quickly and move on.

Many men scrutinize their erections. If you're a man, don't worry about what you see, as every man's erection is a little different. Small kinks or curves (up, down, or in any other direction) are normal. Furthermore, there's a tremendous variability in the angle of a man's erection, depending on the tension of the ligament (page 97) that holds it in position.

Ready to pull out your protractor? The following table shows you the frequency of different penis positions. Zero degrees means your member points straight up, 90 degrees means it's horizontal, and 180 degrees means it's pointing at your feet.

Erection Angle (in degrees, measured down from the abdomen)	Percentage of Men
0–30 (straight up)	5%
30–60 (upward)	30%
60–85 (slightly aloft)	30%
85–95 (straight out)	10%
95–120 (downward)	20%
120–185 (straight down)	5%

The average man has four to eight spontaneous erections each night in his sleep. Doctors sometimes take advantage of this fact to help diagnose patients who have trouble becoming erect during sex (see the box on erectile dysfunction on the next page). The basic idea is to fit the patient's penis with a sensor that tracks his nightly erections. If those erections still take place, the patient's dysfunction is more likely mental than physical.

Penises aren't the only part of the male anatomy that becomes engorged during arousal. The testicles also swell with blood, although the effect is less noticeable. If a man remains aroused for a long time without ejaculation, he may experience a minor condition, known as *blue balls*, which is characterized by a lingering ache or a feeling of heaviness in the testicles. The condition is temporary, and ejaculation (with or without a partner) brings the speediest relief. Incidentally, women can suffer from the same phenomenon, experiencing pelvic heaviness and aching if they don't reach orgasm.

Erectile Dysfunction

No discussion of erections would be complete without mentioning *erectile dysfunction*, the often-embarrassing situation where a man can't achieve a firm erection or maintain it long enough to complete a satisfying sexual experience. While the ravages of time gradually erode the speed, strength, and duration of any man's erection, erectile dysfunction is a chronic problem that can sideline a man's sex life altogether. Possible triggers include psychological causes (such as anxiety and depression), drugs (including alcohol and sedatives), circulatory problems (such as diabetes and high blood pressure), and hormone imbalances (such as a lack of testosterone). If you begin to suffer the effects of erectile dysfunction, swallow your pride and discuss it with your doctor, because there might be more than your manhood at stake.

If you have erectile dysfunction due to advanced age or a known disease, you may choose to treat it with a drug, like the ever-popular Viagra. Viagra relaxes the muscles in the artery walls that lead into the penis, making it easier for blood to flow in. However, Viagra is far from a cure-all, and doctors tell the same story again and again: Viagra works as intended, but people discover their relationship problems have deeper roots.

Incidentally, Viagra has captured the interest of many young, healthy men without erectile dysfunction who hope to boost their sexual prowess—against the drug maker's instructions, needless to say. The results vary from almost unnoticeable (Viagra doesn't *cause* erections, it just makes it easier to have them) to slightly uncomfortable (making the penis feel a tad overinflated). In any case, it's important to remember that very few women ask for a stiffer man or hour-long intercourse. Women are more likely to complain that they're getting short shrift with the foreplay that precedes the main event.

The Matter of Size

It's the source of much locker-room anxiety. And despite the fact that the vast majority of women find no fault with the size of their partner's equipment, over a third of all men suffer from what urologists call *small penis syndrome*—the psychological fear that they just don't measure up down below. And in case you doubt the universality of this fear, consider a type of mass hysteria called *genital retraction syndrome*, a phenomenon that periodically sweeps across some Asian and African cultures. When it occurs, men become paralyzed by the fear that their penises are shrinking or retracting into their abdomens. (To learn about the history of these panics, check out *http://en.wikipedia.org/wiki/Penis_panic*.)

The subject of penis size is complicated by the fact that most men have no idea what the average size is. Rare record setters have topped 13 inches. And an unlucky few suffer from the cruelly named *micropenis*, which is characterized by a dangler that's less than 2.75 inches erect. (Doctors usually diagnose this problem in infants, and they can treat it with growth hormones.) But most men fall into a remarkably narrow range and measure somewhere between 5.5 to 6.2 inches long.

Flaccid penises have more variability. Urologists separate men into two groups. *Growers'* penises start small and more than double their size during an erection. *Showers* are bigger to begin with, but have a less dramatic increase in size when erect. In other words, the erection is the great equalizer among men.

> **Note** No magic pill can increase your penis size. And doctors don't recommend surgical techniques for such a delicate part of your body because they can cause serious complications. If you're not above using deception, there are a few slightly desperate tactics to try—trim your pubic hair so that your penis stands out more, or pad your underwear with a zucchini.

If you're a man who needs an ego boost, picture an adult silverback gorilla, weighing in at a massive 400 pounds of quiet, hulking strength. Would it surprise you to learn that his erect penis is a scant inch long and not much thicker than a pencil? Chimpanzees are in the same situation—better endowed than the massive gorilla, but still far smaller than the average human. In fact, humans distinguish themselves in the animal kingdom with what is, proportionately speaking, the biggest package around.

The reason for this difference is still a mystery, but most biologists believe that humans have bigger penises because, at some point in our history, females wanted them that way. In other words, women may have picked the more impressively outfitted men—not because big penises delivered a better sensual experience, but because, as a secondary sex characteristic, women equated them with health and fertility. (See page 245 to learn why a big member often doesn't stimulate much more than a woman's imagination.) Another possibility is that big penises deposit sperm farther up the reproductive tract, making it more likely to meet with an egg.

Ejaculation

As you can see in the figure on page 237, two passages converge in the *urethra*—the narrow tube that leads out of the penis. One carries urine from the bladder, the other transports sperm from the testicles. If you put all romance aside, from a reproductive perspective the human penis is little more than an overhyped sperm-delivery system.

On its way out, the sperm mixes with fluid from the prostate gland and seminal vesicles to create a thick, nutrient-rich mixture called *semen*. The purpose of semen is to carry sperm safely on its journey to a receptive uterus, where it can get busy making a baby. Toward that end, semen includes sugars, amino acids, vitamins, and minerals, all of which work to nourish and protect the sperm it carries. Semen also neutralizes acids, including those from leftover urine in the urethra and those that deter bacteria in a woman's reproductive tract.

> **Note** Some researchers have even suggested the far-fetched idea that the minute traces of hormones in semen can have a positive, mood-altering effect for the females who receive them. If nothing else, it's a solid basis for some bad pickup lines.

Semen researchers offer some more facts you probably don't want to know about the wonder fluid:

- It ranges from opaque white to yellowish and often has the faint odor of bleach.

- The amount of semen a man releases in one go varies widely based on hydration, arousal, recent sexual activity, age, and personal variation. Studies show that, on average, men ejaculate just 1 teaspoon of semen, although far more is possible. (Presumably it helps to leave the laboratory for a more romantic setting.)

- The volume of semen doesn't reflect the number of sperm present.

- Finally, for those who need to know: Semen is definitely not fattening, with a measly 5 calories per teaspoonful.

Naturally Monogamous: The Tale of the Testicles

When it comes to penis size, humans have the most to brag about in the locker room of the animal kingdom. But that's not the case when the comparison turns to testicles. Chimpanzees carry testicles that dwarf ours and produce prodigious amounts of sperm. Pound for pound, rat testicles are even more enormous—proportionately endowed humans would feel like they were carrying a pair of cantaloupes between their legs.

These comparisons are certainly strange—but do they have any bearing on your life as a human being? Surprisingly, this bit of biology just might provide a few clues about whether you can trust a man to stay faithful.

Here's why: First, animals with proportionately large testicles—like chimpanzees and rats—have them because they need to produce more sperm. They need more sperm because they're in close competition with other males. For example, a female chimpanzee in heat may enjoy the amorous affections of several suitors in a single day. The chimpanzee that produces the most sperm has the best odds of winning the race to the egg. (In fact, sperm competition is a bit more complex than that. The reproductive tract of a female chimpanzee is the scene of a vicious sperm war and the vast majority of chimpanzee sperm—some 99 percent—isn't even able to fertilize an egg. Instead, it's there to destroy the sperm of other males.)

The bottom line? The size of our testicles suggests that humans don't sleep around as much as chimps, and that we're more suited to long-lasting relationships. However, our testicle size still outweighs that of our other cousins, the gorillas and orangutans, which implies that these apes make for more faithful lovers.

Female Sex Organs

Unlike male genitalia (which hang out in the open, slightly desperate for attention), female sex organs are tucked discreetly inside. In the days before modern biology and call-in sex shows, this left most men (and many women) in the dark about exactly what women's bodies *did* in there. It also fueled the still-enduring cultural story that women are the more mysterious sex.

Fortunately, a little biology can clear the fog. Now that you've toured the male sexual organs, it's time to cross the aisle and look down the pants of the other sex.

The Inside Job

The female sex organs do their work in relative privacy. To see what's really happening, you need to take a trip inside.

The first discovery you'll make is that, despite looking dramatically different, the sexual organs of women have a lot in common with those of men. For example, women match men's testicles, which produce sperm and hormones like testosterone, with the *ovaries*, which produce eggs and hormones like *estrogen*.

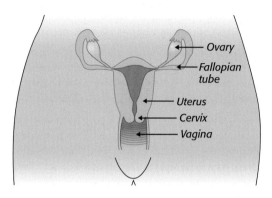

Note Oddly enough, eggs do just fine in the stifling heat of the body, which means that, unlike men, women don't need a dangly and thoroughly inconvenient scrotum to keep their sex cells cool.

Of course, there are some hefty differences, too. For example, ovaries operate on a very relaxed schedule. Even though women have two of them, they produce and release just a single egg each month. (Compare that with the testicles, which produce hundreds of sperm cells *a second*.) Ordinarily, the ovaries take turns producing eggs, although studies suggest that one ovary can do the work of two if the other is removed. This light schedule reduces the chance that a woman will become impregnated with multiple babies at the same time—an event that would have presented life-threatening risks in the not-so-distant past.

Once released, the egg drifts downstream through a narrow passageway called the *fallopian tube*, eventually reaching the *uterus*, which is the egg's final stop on its journey. If a partner's sperm fertilizes the egg along the way, the egg implants itself into the wall of the uterus and begins the long transformation into a new person. Otherwise, the woman sheds the egg within days.

Female Fertility

Women are born with all their eggs in two baskets—their ovaries.

If you're a woman, you begin life with 2 million immature egg cells, and you still have almost half a million when you hit puberty. With each menstrual cycle, your body kills off about a thousand, and releases just one as a potential candidate for fertilization. Interestingly, this process of immature egg destruction continues even if you're pregnant or you've interrupted ovulation with birth control pills. All in all, you have about 400 chances to conceive before you run out of eggs, your hormones shift, or menopause begins. (Typically, menstruation begins around age 12, and menopause sets in between the ages of 48 and 55.)

Clearly, 400 is a staggeringly small fraction of the original millions, which makes scientists wonder if the egg maturation process somehow selects the healthiest candidates. One thing is sure—eggs have to stick around for many years before being used, and that time can cause some of them to degrade. That's why women past the still-young age of 35 may have more trouble conceiving. Similarly, every year of increased age places women at a greater (but still small) risk of producing a child with chromosomal abnormalities, such as Down's syndrome.

The gateway to the female reproductive system is the muscular canal known as the *vagina*. Although its length varies from one woman to the next, the average vagina is a mere 3 inches long. And although no one wants to reduce the mysteries of human sexuality to rulers and tape measures, this fact raises an obvious question: How can a penis with an average erection size of about 6 inches fit inside an average vagina that's only half as long? It turns out that the vagina performs a size-expanding trick that's just as remarkable as a man's erection, albeit less obvious. If a woman is aroused and her partner introduces his member gently and gradually, her vagina stretches to up to twice its length to accommodate him.

This brings up an age-old question. As you've already learned, most women claim to be perfectly happy with partners who have average or smaller penis sizes. Some women are intimidated by larger sizes, while a few prefer them. But how much difference does size actually make in the heat of the moment? From a pure sensation point of view, what matters is penis *girth*, because a thicker member can produce the stretching sensation that some women enjoy. The length doesn't make nearly as much difference because the vagina has virtually no nerve endings inside. Of course, that shouldn't stop women from setting their own personal preferences—after all, men have more than a few fantasies that aren't based on biology.

Can You Spot a Virgin?

There's one more detail between the female sexual organs and the outside world: the *hymen*, a thin membrane that partially covers the vaginal opening. The hymen is best known as the mark of a virgin, as first-time sexual intercourse tears the hymen and often causes bleeding.

But it turns out that the hymen isn't the clear-cut signal it's made out to be. It commonly breaks in childhood for a variety of reasons, from ballet exercises to horseback riding. Some women are born without a hymen, and others have a naturally elastic hymen that stretches open without breaking, bleeding, or causing much pain.

All of these details can cause serious relationship trouble in cultures that prize virginity and have antiquated rituals, like displaying blood-marked bed sheets after the first night of marital bliss. In fact, some non-virgins become desperate enough to turn to surgery before marriage. They may opt for *hymenorraphy*, an operation that recreates the hymen by piecing together its remnants. Another surgical technique involves inserting a gelatin capsule filled with a bloodlike substance, which bursts during intercourse and simulates bleeding. Needless to say, in cultures that demand proof of a new bride's virginity, both tactics are illegal.

Incidentally, the biological purpose of the hymen is something of a mystery. Some believe that it was important in the distant past for sanitary reasons—for example, it may have prevented dirt and bacteria from making their way into the vagina during early childhood. Others believe that the hymen is simply an evolutionary leftover with no true purpose, other than the ones we impose on it.

Pleasure and the Clitoris

In a man's body, a single wonder-organ takes care of several tasks (insemination, urination, and pleasure). In women, three different body parts perform the same duties. The vagina takes sperm on the road to fertilization. The urinary tract is close to the vagina, but separate. And the pleasure center is another nearby neighbor—the pea-sized bundle of tissue called the *clitoris*.

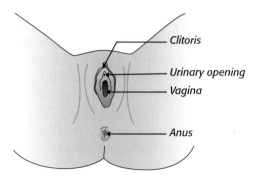

Clitoris

Urinary opening

Vagina

Anus

The clitoris holds the densest concentration of nerves found anywhere on the surface of the human body. It's covered by a hood of skin, which is similar to the foreskin that covers the end of a penis. In fact, as far as pleasure is concerned, the clitoris is a bit like a highly miniaturized, more potent version of the penis. And like the penis, it swells with blood during arousal, becoming erect.

> **Note** Mammals are the only animals that have a clitoris. For most, the clitoris is a clear benefit. Spotted hyenas, however, aren't so lucky—they give birth through their narrow clitorises in an excruciating process that's best left undescribed.

Although there are some women who can bring themselves to orgasm through breast stimulation, foot stimulation, or even—in very rare cases—just thinking hard, the vast majority of women can't get there without the help of the clitoris. If they experience an orgasm without the direct stimulation of the clitoris (in other words, when the penis is hard at work but the fingers aren't), they still have the clitoris to thank. That's because its tissue extends into the body, which allows it be stimulated indirectly through the vaginal wall.

> **Note** Although female anatomy hasn't changed in thousands of years, science has repeatedly discovered and rediscovered the clitoris. With modern biology, the clitoris is no longer a secret—in fact, it's soared in popularity. However, single-minded men beware: The clitoris isn't a magic button. The real source of sexual pleasure isn't any single piece of reproductive anatomy—it's the experience of the woman in question. To get the best advice about what feels good (and what feels so good it's bad), pay attention to her signals, go gently, and if in doubt, ask.

Breasts

We all know their most obvious use—feeding the next generation. But as many a breastfeeding mother will report, it's sometimes hard to put them to that purpose while living in the most breast-obsessed society in human history.

Do I Have a G Spot?

Although the clitoris is the sexual pleasure center, there's another trendy area that gets a lot of attention, despite the fact that it's never been confirmed by science or recorded in a single gynecological textbook. This is the fabled *G-spot*, an area that is thought to reside on the front wall of the vagina. When stimulated, the G-spot may provide intense and slightly unusual feelings. Pleasure may be accompanied by a sensation that feels like a sudden need to pee. Ignore this feeling for a few seconds and it will (hopefully) be replaced by a gradual feeling of sexual excitement—and maybe even a different kind of orgasm.

To find the G-spot, you need the help of a willing partner. Intercourse isn't the best bet, because the area is difficult to reach by all but the most acrobatic lovers. Instead, your partner will need to use a finger to get there. If you're feeling experimental, follow the steps at *www.netdoctor.co.uk/sexandrelationships/gspot.htm*.

As a supply of infant food, human breasts are a small miracle. Beyond the obvious components—fat, proteins, carbohydrates, vitamins, and minerals—breast milk is a cocktail of hundreds of biological compounds like antibodies, immune-signaling cells, natural painkillers (for a short period of time after birth), and even beneficial bacteria.

Note The real surprise when it comes to breastfeeding is the fact that babies who are fed formula, which lacks all the specialized compounds of breast milk, develop just as well as breastfed babies. Yes, Studies show that breastfed babies are better at fighting infections and avoiding allergies, and some suggest more controversial (and relatively minor) links to life-long health and higher IQs. But beyond these relatively subtle relationships, happy formula-fed babies flourish right along with their breastfed counterparts.

Although breasts exist, first and foremost, to nourish babies, just about everything we do with them—from foreplay to cosmetic surgery—is tied to their role as a secondary sexual characteristic. Quite simply, breasts provide a signal that advertises a fertile, sexually healthy female—and that signal almost totally overwhelms their primary purpose. Flip through any night of network television and you'll find plenty of examples of breasts in search of male attention, and very few examples of breasts at work as portable baby bottles. And of course, breasts (or more precisely, their nipples), also serve as potent pleasure centers during amorous encounters.

All of this raises an obvious question: Why are so many people so crazy about pouches of milk ducts wrapped in fatty tissue? Theories abound about whether breast obsession is a natural, gene-powered compulsion, a Freudian holdover from childhood, or a cultural fetish like the ones that give charged meaning to lipstick and long hair. Zoologists point out that breasts provide another example of how humans out-weird the rest of the animal kingdom with their focus on sexual bits. First, human breasts are larger, proportionately speaking, than those of other animals. Second, human females are unique in having noticeable, swollen breasts even when they aren't lactating. These details suggest that breasts have played an important role as attention-grabbers throughout our history.

> **Note** Despite obvious differences, both men and women have one clear thing in common: feelings of anatomical inadequacy. In fact, the large number of women who feel their chests don't measure up matches the large number of men who feel their packages come up short. Similarly, the large number of men who report satisfaction with their partner's breast size closely parallels the large number of women who are happy with their partner's penis length.

The Uterus

The most impressive reproductive organ of either sex is the *uterus*, the pear-shaped chamber where every human spends his or her first days. Not only is the uterus the perfect home for incubating new life, it's also lined with muscle that can, during birth, exert more pound-for-pound force than any other part of the body.

As you no doubt know, the uterus sheds its lining in a regular cycle that runs roughly once a month. This occurrence is uniquely human—most other animals reabsorb their uterine lining rather than discard it, and they have minimal bleeding.

> **Note** Birth control pills suppress menstruation. To shed their uterine lining in the familiar once-a-month cycle, women need a brief hormone-free pause. For years, some women have cheated the system (you know who you are) by taking birth control pills continuously, avoiding menstrual bleeding altogether. Doctors are divided about whether this is a harmless practice or a potentially dangerous change to the body's natural function. Either way, pharmaceutical companies have recently developed a birth control pill, called Lybrel, intended for just this sort of continuous use.

Shortly after menstruation ends, *ovulation* occurs. If you're a woman, this is when one of your ovaries releases an egg. This egg then drifts down to your uterus in a trip that lasts less than 48 hours. If the egg isn't fertilized before this journey ends, it sinks into the lining of your uterus and disintegrates.

Sperm lasts a bit longer in the uterus—usually 2 or 3 days, but particularly hardy specimens can make it up to 5 days. For that reason, a woman's potential fertile period begins 5 days before ovulation (because any sperm that arrive at this time might still be kicking around when the egg finally makes its appearance) and continues until 2 days after (when the unfertilized egg is absorbed).

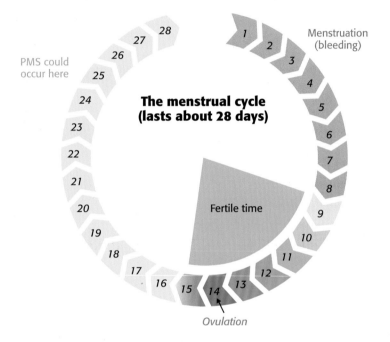

PMS could occur here

Menstruation (bleeding)

The menstrual cycle (lasts about 28 days)

Fertile time

Ovulation

Although there's little you can do to increase your fertility (outside of medical intervention), you have the best odds of conceiving if you fill the 7 days around ovulation with lovemaking.

Tip If the time between your periods lasts longer than 28 days, count 2 weeks back from the ending date to estimate your ovulation date. For example, if you have regular, 30-day cycles, you're most likely to ovulate on day 16 (because 30 – 14 = 16). You can also try to pinpoint the day you ovulate using other signs. For example, when you ovulate, your vaginal discharge increases and becomes thinner and clearer. You may also feel a dull abdominal ache called *ovulation pain*, which typically lasts a few hours (although not all women have this pain).

Vasectomy vs. Tubal Ligation: Who's Making the Trip

Pregnancy may be a miracle of the human body, but you might not want the outcome of every romantic hookup to result in 4 a.m. feedings and foul-smelling diapers. Birth control is the obvious solution, but there comes a time in many relationships when it's time to close up shop for good.

Modern medicine has the answer—in fact, it has two answers, one for women and one for men. Both have failure rates well under 1 percent, and neither has any effect on sexual pleasure or hormone levels. For extra safety, you can do both, but most couples pick one of the two. Assuming you don't want to decide the matter with a coin toss, the following table pinpoints the differences.

	Vasectomy	Tubal Ligation
For	Men	Women
Snips	The vas deferens, the tubes that carry sperm from the testicles to the urethra (page 237). Incidentally, this operation doesn't change the amount of semen that's released during ejaculation.	The fallopian tubes, which carry eggs from the ovaries to the uterus
Type of surgery	A simple, 10-minute procedure in a doctor's office	A more complex, 30-minute surgery in a hospital, with several hours of in-hospital recovery time
Anesthetic	Local (the doctor numbs only the relevant area)	General (you're unconscious)
Effective	As soon as the leftover sperm is gone (typically a month, or 20 ejaculations, later). A follow-up visit can confirm that you're sterile.	Immediately
The aftermath	You may have minor swelling or bruising, which you can relieve with an ice pack or over-the-counter pain medication. The typical recovery time is a day or two.	You may experience abdominal pain, fatigue, and a bloated feeling. The typical recovery time is less than a week. The risk of major complications is very small, but still significantly greater than for a vasectomy.

We extend our apologies to the men. Based on an honest comparison, it looks like the simplest solution is to give them the snip.

The Main Event

Now that you have a firm understanding of male and female anatomy under your belt, you're ready to consider the remarkable biological act we use it for: *sex*.

> **Note** Regular sex is a win-win equation. Various studies suggest that it reduces anxiety and boosts immunity. Most impressively, people who have regular sex appear to react better to stressful situations—their blood pressure rises less and returns to normal more quickly.

Arousal and the Art of Foreplay

When we think of sex, most people jump straight to the anatomical exchange. That's the part where bodies embrace and genitals fit snugly together. But most sex researchers agree that the best sexual experiences begin with a period of biological "warming up"—the stroking, kissing, talking, and teasing called *foreplay*.

Frequently Asked Question

Who Invented the Kiss?

One of the most popular forms of foreplay is *kissing*. Yet even the sharpest scientist doesn't know why we do it.

Some call it an instinct and point to kissing-like behavior in other animals (which is usually done as a part of grooming or to smell one another). Others believe it all started with the intimate act of *premastication*, where a mother chews food and then delivers it to her infant lip-to-lip, which was a necessary skill in the days before baby spoons and infant purees. Still more researchers believe it's a cultural practice that's gained popularity through time. They point out that men used lip-to-lip kissing as a greeting in the not-so-distant past, and non-Western cultures were particularly slow to jump on the kissing bandwagon. (Many felt that the exchange of saliva was revoltingly unhygienic. Experts agree, and report that thousands of bacteria can leap between mouths in a single kiss.)

Whatever the answer, kissing is a lynchpin of modern relationships. Kiss scientists (otherwise known as philematologists) call kissing a point of major escalation or de-escalation. In other words, locking lips may deliver a preliminary verdict on a new relationship, driving it forward or sealing its end.

Pop psychology often describes men as overexcited sex beasts with hair-trigger arousal. It argues that men can plunge straight into intercourse, while women need a period of foreplay to gently build intimacy. But studies that stare straight at your sex organs reveal a different reality. One notable study used a thermal-imaging camera to track blood flow to the genitals as volunteers watched erotic movies. They found that both men and women take essentially the same amount of time to become fully aroused and ready for intercourse—about 10 minutes.

This doesn't mean that women don't want more foreplay than men—it simply suggests that the reason isn't anatomical. After all, part of the purpose of foreplay is to lower psychological inhibitions and increase emotional comfort. The average woman may rely on this effect more heavily than the typical man. Furthermore, men and women often respond to different erotic stimuli. For example, studies find that many women are surprisingly unimpressed by male nudity. The sight of a strapping lad strolling about shirtless is much less likely to cause a jolt of female excitement than a similarly underdressed woman wandering past a male onlooker (which is why amorous men invented roses and chocolate). Studies show that women are more likely to respond to the sights and sounds of sex, even if it's between two women or between wild animals—say, excited monkeys on nature programs. All this adds up to the possibility that an ordinary man might have polished off his 10 minutes of arousal time before the woman has seen anything worth getting excited about.

Note Don't bother trying to shortcut the process with a basket of oysters or a tiger penis. So-called aphrodisiacs, while popular the world over, have never been linked to anything more than the placebo effect.

Orgasms

A successful sexual encounter almost always ends with a burst of pleasure known as the *orgasm.* The experience is brief—rarely longer than 3 to 15 seconds. During an orgasm, your brain releases a batch of pleasure-mediating chemicals and your muscles contract—most importantly in the uterus, which sucks any available sperm deeper inside. However, men take note: Although orgasms and ejaculation are usually part of the same experience, the orgasm is actually a moment of mental pleasure that precedes the physical act of ejaculation.

Shortly after an orgasm, brain chemicals trigger a feeling of relaxation that may serve two evolutionary purposes. First, it keeps lovers lying around for a moment which increases the odds of conception. Second, it boosts bonding, preparing the amorous couple for the next stage in their reproductive journey—parenthood.

Most men experience a *refractory period* immediately after an orgasm. They lose their erection as their sexual parts briefly close up shop. The length of the refractory period varies from man to man, and it grows longer with age, but an enforced 15-minute sex break is common. Many sexperts believe that men can have multiple orgasms in quick succession (like some women can). The trick to experiencing this phenomenon is to pause just before ejaculation.

The Practical Side of Body Science

Reaching the Big O

Despite your best intentions, you may find yourself in an awkward sexual position—aroused, engaged, but somewhere short of the big event. The problem is particularly common in women, but it can affect anyone. Here's some helpful advice for dealing with an elusive orgasm:

- **Focus on experience, not execution.** If you're preoccupied with your performance rather than your experience, you're likely to face frustration. Instead, focus on the experience, breathe, and let go.

- **Get the right stimulation.** It sounds a bit clinical, but urologists agree—when a woman doesn't reach orgasm, it's usually because she's not getting continuous clitoral stimulation (page 246).

- **Beware of social lubricants.** No, we don't mean the oily stuff. Instead, the concern is mood-altering substances, like cocktails, that you may use to help courting run smoothly. The problem is that, if you make it to the bedroom, a body full of alcoholic beverages can thwart the best intentions.

- **It's OK to take a pass.** An orgasm is one component of a lovemaking session, not its final verdict. If pursuing the end event creates more stress than pleasure, or you're exhausted and in need of a break, try again tomorrow.

Toward Better Sex

You've now picked up enough trivia about the male and female reproductive systems to shock the most experienced serial dater. But one question remains: Will any of this help you become a better lover?

Unsurprisingly, there's more to sex than being able to describe the content of semen. Being comfortable with your body (and knowing your way around your partner's equipment) establishes a good foundation for a life of healthy sex and reproduction. And if you still feel like you need a little something more, try these tips:

- **Meet your most sexual organ.** It's your mouth. Not for the reasons you might think, but because *talking* is the best way to discover your partner's secret fantasies.

- **Engage in mutual exploration.** Novelty is a fuel that keeps sexual life burning. If you're interested in trying new sexual positions, you can browse through an encyclopedic list at *http://en.wikipedia.org/wiki/List_of_sex_positions*, see 3-D models at *www.sexinfo101.com/sp_index.shtml*, or kick it old-school with the Kama Sutra.

> **Tip** Remember to follow the first point: For sexual experimentation to work, it needs to be a shared, trust-building experience. The pleasure is in the sense of shared exploration, not in any anatomical magic. People who forget this run into trouble with the first *Cosmo* article they read. (In fact, independent studies find that many of the infamous sex tips in magazines like *Cosmo* and *Men's Health* are surprisingly unpopular, particularly when launched, unadvertised, on an unknowing partner. Notable non-starters included nipple pinching, unexpected toe sucking, and mixing peppermint candies with oral sex.)

- **Express yourself.** Sex is more than what goes on between two people's genitals. Women (and men) want more than just mechanical stimulation—they need to be desired, craved, and hungered after.

- **Start training.** It's a bit ambitious, but some people swear by the sex-boosting effects of *Kegel exercises*, which strengthen your pelvic floor muscles (these are the ones you feel when you stop urinating midstream). Strong Kegel muscles let women clench their vaginas, and may give men more control over erections and ejaculation. If this doesn't already seem like way too much work, you can learn more at *http://tinyurl.com/cqtwln*.

- **Don't judge.** Before you write off a sexual practice as unnatural, consider the fact that plenty of animals are probably already doing it, far from the corrupting influence of human society. Masturbation, same-sex action, orgies, necrophilia, and bondage are a few of the sexual taboos being happily (and frequently) violated in the animal kingdom. To see some famous examples, you can read a two-part exploration in the magazine *New Scientist* at *http://tinyurl.com/d46wky*.

11 Your Final Exit: Aging and Death

I n the natural world, far from human society, life is nasty, brutish, and short.

We alone, of all species, have taken ourselves out of that environment and built a comparatively easygoing playground to live in. In our modern societies, we enjoy lives of relative contentment, free from natural predators and the daily fight for survival. The only wrinkle in this peaceful world is *death*—it keeps happening. Although death has always been part of life, now it happens with a twist. Today, death is less likely to come from a fight with a rival or a misadventure in the wild, and more likely to result from the slow erosion of your body's operating hardware—a system-wide failure that gradually takes every organ offline.

This phenomenon—death by *old age*—is almost completely unprecedented, and it's a concept that's alien to every other species on the planet. We call it dying of natural causes, stubbornly ignoring the fact that there's nothing natural about it. Being eaten by an alligator? Natural. Death from exposure? Just as ordinary. Consumed by parasites? Slowly starved over a long, bitter winter? Both perfectly understandable in the natural order of the animal kingdom. But to make it through these events to a time when the body simply runs out of steam, to see it succumb without a single external threat, its energy seeping gently away like a leaky balloon, is distinctly *un*natural.

In this chapter, you'll confront this riddle head-on. You'll estimate the years left in your life, consider some of the things that may do you in, and try to turn back the clock. You'll also take a grim journey to the final moment to learn what's in store at the end of your days.

Life Expectancy

It may take 80 years or it may take just 20, but at some point you'll reach the final chapter of your life.

For some, the idea is as chilling as a bucket of ice water in bed on a Sunday morning. For others, the unavoidable, undeniable certainty is almost a comfort. But nearly all of us agree on one point: We want to know how much time we've got left before the end.

Your Lifespan

Estimating your lifespan is a series of guesses. One good starting point is to consider the country you live in. It's little surprise that wealthy Western nations have the highest life expectancies, as they have ample available food, basic disease-prevention measures, and access to life-saving medical intervention. (Other factors that play a role—good genes, a healthy lifestyle, and a positive outlook on life—don't show up on this sort of map, because for every good-living superstar there's a sorry soul with sketchy genes or a bad attitude who's balancing out the average.)

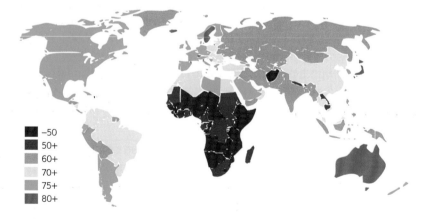

-50
50+
60+
70+
75+
80+

This map tells three stories. People in African countries like Zambia have a low life expectancy—the result, it turns out, of two grim diseases: AIDS and malaria. Avoid these diseases, and you're much more likely to coast into old age. In Asia, survival is particularly difficult in large, rural communities, where access to medicine is poor and infant mortality high. Finally, in countries un-encumbered by these factors, people do best of all. Men can expect to reach age 75 years, while women are likely to slightly surpass 80.

Note In all the countries and cultures of the world, women live longer than men—usually by a 5- to 10-year gap. There's no ironclad explanation, but researchers pinpoint two solid possibilities. First, women tend to develop heart disease many years later than men. Second, young men, pumped up on testosterone, are more likely to engage in decidedly risky activities, like extreme sports, playing with weapons, and driving fast without a seatbelt.

The Ultimate Limit

There's another way to look at the life-expectancy numbers: through time.

At first glance, historic life-expectancy figures appear to show how the advances of modern society have vastly expanded our lifetimes. For example, the life expectancy of a white male born in the U.S. in 1850 stood at a pitiful 38 years, roughly half of today's figure. However, this number is more than a bit misleading. Despite what you might assume, it doesn't mean that the average person in 1850 lived precisely 38 years and then died. The real story is a little more complex. That's because the calculation for average lifespan in 1850 balances two very different groups of people. On one hand, scores of infants and children died at birth and from childhood diseases. On the other hand, large numbers of people lasted into a respectable old age. When you average the ages of death, you end up somewhere between these two groups (at age 38 in the case of the 1850 population).

The best way to understand the significance of childhood mortality is to look at the *remaining life expectancy* for people at different ages. As you already know from the life-expectancy map (page 258), a newborn baby boy in our time should stick around for some 75 years. But how do the odds change if that boy reaches 10, 20, and 30 years old? The following table answers these questions and compares the numbers with the same figures from the distant past.

	Years Left to Live at Age...								
	0	10	20	30	40	50	60	70	80
Distant past (1850)	38	48	40	34	28	22	16	10	6
Current day	76	66	57	47	38	29	21	14	8

This table tells an interesting story. If you make it to your 10th year of life, you can expect to live 66 more years. In other words, your prospects haven't changed from your life expectancy at birth—statisticians still expect you to live to be 76 years old. If you reach age 20, life is looking up just a bit—you'll probably last 57 more years, which means you've added 1 year to your estimated lifespan (20 + 57 = 77).

The situation was dramatically different for a man in 1850. He started with the odds stacked squarely against him, with a life expectancy of just 38 years. But if he made it 10 years, his life expectancy immediately climbed to 58 years (10 + 48 = 58). As the years went by, the gap continued to narrow, so much so that a 60-year-old man living over a hundred years ago was likely to be just 5 years (and one decent pair of dentures) short of the existence he'd enjoy today.

This trip back in time makes a grim point. Medical advances increase the odds that we'll dodge dangerous diseases that would end our lives early, particularly in childhood. But the incredible advances of medicine haven't done much do stretch our *maximum lifespan*, which still stands about 10 years short of the century mark. So the story of longevity over the last hundred years is this: Far more people make it to old age, but old age hasn't gotten much older.

Note For further proof that a typical person's maximum lifespan hasn't changed much, check out old texts. Shakespearean plays and the Bible consistently refer to human life as a period of "three score and ten" years (that's 70 years), a number that's not far from the modern mark.

Hoping to defy the odds and live really, really long? The oldest person who ever lived (that we know of) reached 122 years. There are no other confirmed cases over age 120. Oddly enough, all but one of the oldest known people living at the time of this writing are 113, and all but one of those are women. You can follow the progress of the top-100 *senior* senior citizens at *http://en.wikipedia.org/wiki/List_of_Living_Supercentenarians*.

Enduring Your Long Life

Studies show that the gap between life expectancy and *healthy* life expectancy is gradually widening. The average person today can expect to cap off a happy life with a full 10 years of ill health.

The problem is that while modern medicine boosts the odds of surviving disease, it hasn't developed a way to halt aging. So as we linger on into old age, we gradually suffer the accumulated affects of our physical decline. Diseases that would have been life-ending disasters in the past are now chronic conditions that can cause constant pain, deafness, blindness, and immobility.

Which raises a natural question: Is there anything you can do to improve your final decade? The best advice is to follow the health tips you've learned throughout this book to control fat, fight disease, clean your lungs, strengthen your heart, and maintain your muscles and bones. Though these practices might not add a single extra year to your life, they will almost certainly influence the *quality* of your old age, allowing you to stay engaged with the world and giving you the use of your body for many years to come.

Why We Die

The revolutions of modern medicine (and modern sanitation) have barely nudged the maximum human lifespan forward. So it's clear that we're up against a formidable obstacle. Or, to put it another way, humans have successfully refashioned the planet, created an arsenal of high-powered pharmaceuticals, sent space probes to distant planets, leveled mountains, constructed islands, pruned entire branches' worth of animals from the tree of life, and probably even made the temperature climb across the globe. But we've added mere years to the upper limit of our lifetimes. Clearly, living longer is not going to be easy.

The Naturalness of Death

To understand how you can increase the odds of extending your life, you need to know why they're so decisively limited by death. After all, it's easy to understand what happens when you meet a hungry tiger in a dark alley, or take the wrong step off a narrow bridge. But dying of old age is another thing entirely—not a sudden break, but a gradual buildup of complications as your biological machinery begins to fail. We expect it, and yet few can explain what makes a natural death "natural."

When faced with this riddle, biologists and philosophers have tried several answers:

- **We die to make way for new, stronger people.** In other words, death is designed to improve society. If its most senior members stuck around forever, they'd hog all the food and resources, stifling the growth of the next generation. (Yes, this accurately describes the worldview of most teenagers.) Unfortunately, this tidy answer is almost certainly wrong. The key problem is that evolution filters out the most successful individuals—it doesn't promote the health of a whole species. So a long-lived, super-successful mutant wouldn't harm his species. He'd take over.

  Verdict: False

- **Death represents a fundamental limit.** In other words, it's a technicality—life just doesn't have enough tricks to keep people alive beyond a certain maximum span of time. This answer describes how we feel, but its science doesn't measure up. First, it defies the many miraculous things human beings can do. For example, we can (in partnership), combine two cells to build a completely new, perfectly functioning being—a feat that's surely more challenging than maintaining our current bodies. Second, it ignores the animals around us, each of which has its own, highly idiosyncratic lifespan limit. If a giant tortoise can surpass 200 years, why does your pet turtle drop dead before 40? And if a clam can live past 400 years, why does a housefly get just 4 weeks?

  Verdict: False

- **We die because there's no reason to keep us alive.** Evolutionarily speaking, we're of no use once we lose our ability to make babies. To understand why, you need to know a bit more about how evolution works. Basically, evolution is about the fittest creatures—those with the best mix of genetic traits or strange mutations—overwhelming everyone else. This happens because fit individuals have more babies than their less successful counterparts. But humans stop having babies around the half-century mark. So even if they live longer, they can't do much to spread their genes to future generations. The bottom line? With no evolutionary pressure encouraging us to stick around, our modest lifespans haven't improved much over the past 100,000 years.

  Verdict: Promising

There's a potential flaw in the last argument. Assume it's true—we die before 100 because we're unlikely to make large litters of babies in our nineties. What's to stop us from evolving into something even better—namely, a race of sexually tireless seniors?

The problem is that, until very recently, we wouldn't last long enough for it to matter. The vast majority of our cave-dwelling ancestors died as a result of famine, predators, accidents, or extreme weather conditions. Very few had the chance to reach the end of their so-called natural lifespan. In other words, their maximum lifespan wasn't very significant (and it really made no difference to anyone if they could increase it) because the overwhelming majority of people never reached it or got anywhere close.

If you're still puzzling this situation out, think about it from an evolutionary perspective. Imagine that in the deep, distant past, a rare mutation cropped up that extended a person's lifespan by 10 years. Odds are that the mutation wouldn't be an evolutionary advantage because the person would die of some other cause before enjoying the extra decade of old age. The end result is that the rare mutation would remain rare or die out altogether.

If you look at other species, you can find some solid evidence that backs up this argument. Some animals, like rabbits and mice, spawn quickly and prodigiously. They play hard, make lots of babies, and burn out early. In the wild, their fast rate of reproduction is an important life skill—the average mouse will last only a few months before meeting its end at the hands of predators or the environment. For that reason, there's never been enough evolutionary pressure for mice to develop a long maximum lifespan. You can see this if you place a mouse in captivity. There, even though the mouse is free from external danger, it manages a maximum age of just a few years.

By comparison, other animals, like elephants and tortoises, face fewer predators and can afford the luxury of a long lifespan. After all, an adult elephant is big enough to intimidate almost any would-be predator (aside from human poachers), and the tortoise sports a tough shell that discourages all but the most persistent attackers. As a result, these animals usually reproduce more slowly and take more time to raise the next generation. (Similarly, birds—which have the option to fly away from most threats—outlast the average mammal.) Humans do fairly well by comparison—we're close to the long-living and slow-reproducing end of the spectrum.

> **Note** If you're waiting for modern science to lengthen your lifespan, don't hold your breath. (It'll just make you dizzy anyway.) Whether you look at historical life expectancies or compare our lifespan to that of other species, the same fact stands out: We appear to have a stubbornly fixed, upper expiration date of about 120 years.

The Blessing of a Short Life

The sex-and-death theory explains why evolution doesn't encourage us to live longer. However, the situation is probably worse than it first appears—not only does evolution fail to give us longer lives, it can also encourage traits that kill us early.

For example, imagine a rare mutation occurs that speeds up aging, ensuring that you'll die before 60, but giving you a real advantage in your child-rearing years. Odds are this mutation would spread wildly, because the benefit (a certain short-term improvement in the years when you can spread your genes) far outweighs the disadvantage (a decline for the very few that make it to old age). If this sounds like a wild theory, consider a few age-related problems that evolutionary biologists suspect might have hidden benefits:

- **Hemochromatosis.** This relatively common genetic disease causes excess iron absorption and destroys the liver by middle age. However, in an iron-scarce environment, it could help half-starved humans avoid anemia and death.

- **Ulcers.** Some people have a genetic condition that ramps up stomach acid production, making stomach ulcers more common in old age. (And as trivial as they sound in today's world, an untreated stomach ulcer can cause complications and death in old age.) However, it's possible that high levels of stomach acid destroy potentially deadly pathogens that are present in food, helping stave off disease earlier in life.

- **Immune system.** It's a spectacularly effective system for battling bacteria, viruses, and other microscopic attackers. However, the side effects it unleashes, such as chronic inflammation, can damage tissue and encourage problems much later in life, like heart disease.

- **Cancer.** Cancer occurs when a mutated cell runs rampant, colonizing the space around it. Your body has a number of mechanisms to battle cancer, including a built-in limit that restricts the number of times a cell can replicate before it degrades. Malignant cancers need just the right mutation to get around this limit, which makes cancer far less common than it would otherwise be. However, the replication limit might also prevent your body's cells from replenishing themselves, forcing your body to decline gradually as it ages.

> **Note** To learn more about the controversial connections between the diseases that ail us and the reasons they keep kicking around the gene pool, try one of the following thought-provoking books: *Why We Get Sick* (Nesse and Williams) or *Survival of the Sickest* (Moalem).

Starving Your Way to Longer Life

In the struggle to stave off death, some people are trying a punishing tactic—extreme caloric restriction. Usually, this means eating 30 to 50 percent fewer calories than in a normal diet. The hope is that this chronic food restriction will force your body to slow down its metabolism, reducing the rate at which your body ages.

On first reflection, this idea seems to fall somewhere between crackpot fantasy and severe masochism. But there is a thin, tantalizing layer of scientific evidence that it might work. Putting mice on a similarly reduced starvation diet can increase their lifespan by 50 percent. And many other species seem to react in a similar way—possibly even primates like Rhesus monkeys and, well, us.

But before you sign up for a life of self-deprivation, there are a few caveats. First, some studies show that caloric restriction works its magic only if you start early in life. Presumably, early food deprivation flips some genetic switches that can never be set again. Second, most calorically restricted mice shut down their reproductive systems. They don't have babies and they don't bother to mate. Third, simply being thin doesn't cut it—studies show that skinnier-than-average people don't get a magical longevity boost. Extreme deprivation is key. Lastly, caloric restriction can lead to other undesirable side effects, like muscle atrophy, bone loss, and—not surprisingly—out-of-control cravings and a maddening obsession with food.

Aging

The vehicle that transports you from hearty youth to senior citizenship and eventual death is *aging*, a gradual process that corrodes every cell of your body. Even as your mind is accumulating life lessons, personal experiences, and bad puns, your body is steadily rusting from the inside out. By the time you see the signs, the damage has spread throughout your body.

Aging reminds us that each person truly is a mind "fastened to a dying animal" (to quote Yeats). When that animal goes lame, it doesn't matter how sharp you are—you're going to have trouble hobbling forward.

The Mechanisms of Aging

Biologists still debate what causes the profound effects of aging, and why some species burn through them so much faster than others. Many theories suggest that we age because our cells accumulate a lifetime of damage. That damage may be caused by wear and tear or chance events. It could be the handiwork of toxins, free radicals (highly reactive molecules that can damage cells), waste products, inflammation, or DNA mutations—all of which exert a minimal effect early in life but gradually become more serious.

Other theories suggest that aging is at least partly a genetic program that's hardwired into our cells. As a person ages, different genes come into play. Aging cells lose the ability to reproduce and replace themselves, possibly limited by a built-in cellular safeguard that prevents cancer.

What *is* certain is that aging affects cells, processes, and organs throughout your body. If you avoid one possible source of damage—for example, the DNA-scrambling rays of the sun or the artery-inflaming compounds in a bacon sandwich—you'll still suffer many others. And while it's hard to tell which line of attack will eventually do you in, age-related changes are profound and far-reaching. They affect everything from the skin that coats your body to the internal organs that sustain it.

The physical decline of aging begins gently in your late twenties and early thirties. Most people first notice the aging process through undeniable changes in their hair and skin. At the same time, similar irreversible shifts take place inside your body. Your heart ages, growing thick and weak. Circulation slows and arteries stiffen. Lung tissue loses its elasticity, reducing the amount of air you can suck in with each breath. The brain shrinks, the kidneys shrivel, and your ability to touch, smell, and taste decline. The ability to hear high-pitched sounds, distinguish fine gradations of color, perceive depth, and spot something in your peripheral vision all diminish. Your body begins to break down your bones faster than it rebuilds them. Everything slows, from digestion to reaction time. It's a long trip down.

> **Note** While we have yet to discover a person who ages more slowly than normal, science has found some who appear to age much more rapidly. These are victims of *progeria*, an extremely rare condition where a genetic mutation disrupts part of normal cellular functions. The result resembles rapid aging, and sufferers die on the brink of adolescence.

Staying Young

Aging is good for one thing: It makes a great number of people a great deal of money. To make your fortune, you simply need to convince others that you have a drug, treatment, lifestyle practice, or magic bracelet that can counteract the processes of aging. And while most of us realize that you can't reverse the clock, many let themselves believe that they just might slow it down a few fractions of a second.

Anti-aging products are tremendously popular with people on the cusp of aging-related losses in health and appearance, such as individuals in their thirties, forties, and fifties. They get far less attention from individuals in their seventies, eighties, and nineties, who've lived long enough to understand that aging is a natural, far-reaching process that pays little attention to our attempts to challenge it. The truth is, it's easy to be youthful when you're young. But as time passes, aging ravages the human body as surely and irrefutably as the Colorado River scours the Grand Canyon.

Faced with this reality, is there anything you can do to bargain for a few more years or at least enjoy them in better health? While there may be no real way to resist aging, there are certainly a number of ways to compound its effects. (As the Renaissance poet Francis Quarles famously observed, "It lies in the power of man, either permissively to hasten or actively to shorten, but not to lengthen or extend the limits of his natural life.") So rather than attempt to deny the inevitable, why not ask yourself how you can accelerate aging—and then do exactly the opposite. Here's some bad advice that can steer you in the right direction:

- **Load up on toxins.** They damage cells and can trigger the broad, functional changes that are the start of many chronic diseases. Poisons like lead and asbestos are obvious examples, but excessive drinking, smoking, and sun exposure are more popular ways to kick-start the aging process.

- **Cease all physical activity.** It's tiring anyway. If you're in your thirties or beyond, underused muscles will waste away that much quicker. You'll lose your mobility faster, and you'll have a greater chance of turning a commonplace injury into a lingering health concern.

- **Cut yourself off from the outside world.** A life of solitary old age is linked with poor survival rates for many common diseases. Those who remain engaged, stay social, continue traveling, and find a way to force out a little optimism have a nasty habit of sticking around.

- **Eat with wild abandon.** Excess weight compounds a host of age-related diseases. It may be because fat is a physical burden, or it may be that fat triggers inflammation, which can cause further weakening in aging organs. There's a similar story for refined carbohydrates and sugar, which can help jumpstart diabetes.

- **Refuse your medication.** If you suffer from cholesterol or high blood pressure at an advanced age, your body has probably run through a series of subtle one-way changes. Even a perfect diet and an Olympian exercise regimen may not change its course. So if ill health and an early death don't concern you, don't bother to monitor cholesterol and blood pressure with your family doctor, and adopt a relaxed on-again, off-again schedule for taking your medication.

Late-Night Deep Thoughts

Aging with Indignity

In today's world, there's a deep, unquestioned connection between youth and health. Talk-show hosts and health gurus implore us to "live younger" with a collection of health recommendations (exercises, proper nutrition, avoiding toxins) that have just one thing in common—we ignored them all when we actually *were* young. The idea is even fused into our language. In today's vernacular, *looking old* means looking bad, and *looking young* refers to everything but the pimply-faced embarrassments of adolescence.

The situation is a bit odd. You need nothing more than a trip to the local McDonald's to remind yourself that most young people aren't all that interested in being healthy. But the rest of us are desperately interested in being like them, even if it's only through an act of shared pretense that bestows no health benefits at all. Hair dye is an obvious example—few are the brave females who tolerate gray hairs before 60 (if at all). In order to fit in, they pretend to have their original hair pigment, we pretend that this makes perfect sense, and the people selling the hair dye pretend this sort of thing will make us feel so much better about ourselves that we are actually becoming younger and healthier in the process.

Now, there's nothing wrong with a little self-deception. After all, it helps us get through the daily social grind. But there's a danger in linking looking young to being young, and then link being young to being healthy. These equations allow many people to become outrageously rich peddling utter nonsense that sounds reasonable—for example, that exercise will prevent wrinkles and other natural signs of age, or that the best expression of self-love is a bottle of fresh Botox.

What we all need is a little more self-respect. After all, how can we prepare ourselves to face the long road of declining health and eventual death if we're convinced that our true self is the person we were at 24? At some point, we need to recognize that there is a season for everything. Life is a space of limited opportunity, bracketed on one end by the vigor of youth and, on the other, by the quiet reflection of age. And if you want to enjoy the entire thing—even the less comfortable years at the end—you need to give up the hunt for mythical life extenders, stop trying to patch up the changes with increasingly desperate cosmetic measures, and find the dignity to do what few of us want to do—age.

The Anatomy of Death

When we think of death, we usually focus on the last few hours of life. However, death isn't a discrete event. Its foundations are laid shortly after adolescence, when your body begins to age and degrade. The process picks up speed when you develop a chronic condition like high blood pressure, diabetes, or heart disease. At this point, the slow process of dying accelerates as small complications lay the groundwork for the eventual end, which may not follow for decades.

In fact, if you live to old age, there will probably be nothing particularly remarkable about the event that finally tips you over the edge. Throughout your life, your body must continuously fight off challenges from infections, injury, and misbehaving cells. A death of old age is usually at the hands of one of these ordinary attackers, and it's a threat that a younger body would easily dispatch.

Common Ways to Die

There may be just one way to enter this world, but life provides dozens of different exits. In Western countries, where war and infectious disease are unlikely to claim our lives, there are two top killers that stand out from the pack. Heart disease tops the list, causing nearly a third of all deaths. Not far behind are the many variations of cancer. The table shown here lists the 14 most likely causes of death in the U.S.

Percent	Cause
28.5%	Heart disease
22.8%	Cancer
6.7%	Stroke
5.1%	Lung disease
4.4%	Accidents
3.0%	Diabetes
2.7%	Influenza and pneumonia
2.4%	Kidney disease
2.5%	Alzheimer's disease
1.4%	Septic shock
1.3%	Suicide
1.1%	Liver disease
0.7%	Parkinson's disease
0.7%	Murder

It's important to note what this chart doesn't include—namely, death by old age. When someone dies, doctors carefully pinpoint the final coup de grâce. However, it's all a bit arbitrary. Once the ravages of time weaken your body, nearly all the entries in this table become more common, and the lines between them become blurred. Diabetes can cause stroke and kidney failure. Strokes can damage the swallowing reflex that protects your upper respiratory system, leading to pneumonia. Pneumonia can cause inflammation and trigger heart attacks. Or, without a strong immune system, the infection can spread throughout your body, leading to septic shock. And so on. One problem bleeds into the next one, but the real culprit is the accumulation of years. When medical examiners conduct an autopsy of a person who has died at an advanced age—no matter what the cause—they commonly uncover advanced heart disease, additional infections, and incidental cancers. When age is involved, the specific cause of death is often semantic.

How to Die with Dignity

Death is a messy business. Few of us will have the experience of dying peacefully in our sleep. If you avoid acute sources of death—things like accidents and instant-death heart attacks—your end is likely to be slow and awkward.

Faced with this final challenge, it's worth asking what you can do to give yourself the best possible sendoff. Here's what the terminally ill tell us they need:

- **Control.** Death isn't just the end of a life, but its last statement. As people approach death, they often feel a deep need to choose the right ending—one that aligns with who they are. For some, this means choosing the place of death (for example, at home rather than in a hospital). For others, it means having certain people or things present—loved ones, music, poetry, or a really big TV. For a few, it means being able to choose when the final curtain call comes, without resuscitation attempts that only stretch the last few weeks of existence thinner and thinner.

- **Comfort.** Although death itself may not be painful, the days and weeks leading up to it often are. Pain medication, cleanliness, and—in a hospital environment—a sympathetic staff can make the difference between relative comfort and last-minute panic.

- **Closure.** Almost everyone wants a chance to tidy up the loose ends of life before they reach the end. These final acts can include saying goodbye, writing letters, and leaving mementoes. There are also several things that are more difficult to accomplish in the final stages of disease, such as managing financial affairs, planning a funeral, or writing a will. However, these issues often become tremendously significant to dying people. The best advice is to think ahead and deal with them while in good health, long before the threat of death arrives.

- **Honesty.** Death is a difficult experience to share with friends and family. Often, well-meaning practices that might encourage a dying person in better days (claiming recovery is just around the corner, denying a problem is serious, refusing to discuss funeral plans) will only cause increasing alienation.

- **A happy life.** One of the best ways to die a good death is to lay the groundwork now, by consciously choosing to live without regret. That means taking risks and pursuing opportunities, so that when the end comes, you'll have the peace to reflect on what you did rather than the resentment to ponder what you didn't.

Treating the Terminally Ill

Modern medicine is organized around treatments and cures. It's less adept at handling gradual decline and inevitable death. Even when a person suffers from late-stage cancer or advanced dementia, the default stance of modern medicine is often to keep them alive at all costs.

For example, Alzheimer sufferers usually die long after they've lost the ability to speak, walk, feed themselves, and recognize loved ones. The actual cause of death is usually complications related to the disease or opportunistic infections like pneumonia. Despite this grim situation, many family members feel powerless to choose any approach other than the sort of aggressive treatments doctors use to cure healthy people, such as courses of antibiotics. These measures may add weeks or months of life, but the sufferer will not recover.

In this situation, doctors are torn—they themselves often feel that withholding antibiotics is the best approach, but aren't comfortable raising the issue with family members. And when doctors deal with the deaths of their own loved ones, they often face the same dilemma, and guilt and grief traps them into choosing medical interventions and invasive treatments that only prolong the pain.

Facing death is difficult, and there's no clear-cut solution to the issues raised by terminal illness. However, you can prepare yourself (and your family members) to deal with the possibility of a long-term illness by making hard decisions beforehand. Discuss with your loved ones what treatments you want them to pursue if disease has stripped away your quality of life. Recognize that withholding antibiotics or letting the final stages of a disease run its course are acceptable options when there's no longer hope of a cure. Realize that after a loved one dies, relatives often regret the treatment decisions they made to keep that person barely alive. And accept that we must balance two very different goals in our time on Earth, as we struggle for a long life, yet hope for an easy death.

The Final Moment

It's difficult to say what the final moments of your life will be like without knowing what will ultimately do you in. But even though death comes in a thousand flavors, there are some common themes. The most important of these is the link between oxygen and life. No matter how you expire, the key step usually involves a disruption between the supply of oxygen and the cells in your body that crave it. It may be that your lungs can no longer obtain oxygen, your heart no longer pumps it, or a blockage prevents it from reaching your brain and vital organs. No matter the cause, the ultimate result is the same.

Here's how the average death unfolds:

1. **The rapid descent**

   The classic death of old age is preceded by a period of gradual weakening. Your body slows down, and you begin to sleep more to conserve your last dregs of strength. As death nears, you may lose the desire and ability to eat or drink. Your control over your bowels and your bladder may also fail as your digestive system slowly passes offline. Finally, your breathing may become labored and irregular, alternating quick, deep breaths and slow, shallow ones as your brain loses its finely tuned ability to manage the body's levels of oxygen and carbon dioxide.

   Note To take a tour of some more exotic forms of death (from drowning to electrocution), you can read a decidedly morbid *New Scientist* article on the subject at *http://tinyurl.com/o4ev3k*. You'll find that though these varieties of death start out quite differently, they quickly converge with the list on this page, starting at the second or third step.

2. **The agonal phase**

   The final, brief moments of life are called the *agonal phase*. They can include muscle convulsions, heaving gasps, and even a startling bark. Fluid that's built up in your lungs may cause a gurgling sound called the *death rattle*. Although the agonal phase is difficult for others to watch, you're probably too far gone to experience any pain. As the sidebar on page 274 suggests, you're more likely to feel calmness, a sense of distance, and disorientation. Inside, your brain is starved of oxygen, and its neurons begin to die.

3. **The immediate aftermath**

   Within a minute, your heart stops beating and your face acquires the unmistakable gray pallor of death. Throughout your body, cells continue their regular work for a few minutes, while nutrients decline and waste products build up. Finally, the stress of this situation triggers cellular changes and self-destruction. These processes set in within minutes, and after this point resuscitation is impossible.

   Once your heart stops beating, your body begins to cool, losing a degree or two of heat every hour until it reaches room temperature. Within 3 hours, *rigor mortis* begins, as calcium seeps into your muscle fibers, causing them to stiffen. Contrary to popular opinion, hair and nails do not continue to grow after death, although they may look longer as the skin shrivels back.

4. **The start of decay**

A couple of days after your death, the billions of cells that powered your body have broken down. The bacteria that you carried all your life, both on your skin and in your intestines, begin to digest your body from the inside and outside. The result is a foul-smelling gas that bloats your body and an obvious discoloration that changes your skin to green, purple, and then black.

Of course, by this point all that's left of you is a body, and that body is unlikely to be left decaying in the open. Instead, it's disposed of according to its former owner's custom—for example, buried, cremated, or embalmed.

Late-Night Deep Thoughts

Dying: The Experience of a Lifetime

One of the most interesting parts of the death experience is the visions that it often brings. We know about these through survivors of *near-death experiences*—people who've been on the verge of the final exit, but clawed their way back. Examples include people who collapsed with dramatic heart attacks, started drowning, or suffered catastrophic injuries and bleeding.

The most tantalizing part of the near-death experience is that the people who've been through it report strikingly similar experiences. Often, there's a feeling of calm, a warmth, and a sense of floating above reality and looking down at yourself. There may be a loud buzzing or ringing noise, and—more often than not—a rapid trip through a dark tunnel toward a dazzling light. Finally, a near-death experience may include a meeting with angelic beings or dead relatives.

It may all sound rather fanciful, but it's important to note that the experience of near-deathers doesn't feel like a dream. Instead, it has an incandescent clarity that seems more real than ordinary life. And the effect is often transformational. After a near-death experience, near-deathers usually lose their fear of death and gain a new outlook on life and their place in the universe. But they're often afraid to talk about it because their friends think, quite sensibly, that they've gone stark-raving bonkers.

Some suggest that near-death experiences are a preview of the afterlife to come, pointing out the uncanny consistency that near-death experiences have had across time and cultures. Scientists and more level-headed people believe that near-death experiences are the byproducts of a dying brain. They link it to other hallucinatory experiences, like epileptic fits, traumatic events, and the effects of drugs like ketamine. The culprit could be a lack of oxygen in the brain, combined with the release of endorphins (a pain-soothing brain chemical)—but the fit isn't perfect, because these processes rarely produce the sense of deep tranquility that near-deathers describe.

Still puzzled? There's one bit of good news. Wait just a few decades (or less) and you'll get the chance to see for yourself.

The Final Word

You're now at the end of your eye-opening journey through the human body. Along the way, you learned how your body builds skin, fat, muscles, and bones. You peered inside to see essential organs at work, like your heart and lungs. You traveled your body's intricate passageways, exploring the arteries and veins that nourish your cells and the winding digestive tract that carries your last meal away. You watched your body translate light, chemicals, and vibrations of air into a rich medley of sensory experience. You saw it battle disease. And you experienced the pleasures of sex and the pains of death.

It's all a bit overwhelming. Because no matter how intimately we think we know our bodies, their amazing capabilities, fatal flaws, and messy compromises are nothing short of staggering. Hopefully, you've picked up enough practical advice to become a better body owner, whether it's through regular exercise, good nutrition, or disease prevention. But even if you aren't planning to change anything, you've certainly learned how to enjoy your body's quirks and capabilities—and that's all part of the fun of life.

If you're wondering where to go next, you can check out the follow-up reading that's listed, chapter by chapter, on the "Missing CD" page for this book at *www.missingmanuals.com/cds*. It links to websites and books that expand on the themes you've touched on in these chapters. And if you're ready to launch a completely different trip, you may be interested in *Your Brain: The Missing Manual*, a companion book that leads you into the equally amazing world inside your head. In the meantime, kick back and get a good rest—your body has earned it.

Index

Symbols

8/14 rule, **78**
20/20 vision, **110**
911, when to call, **170**

A

ABCDE rule, **17**
abdominal crunches, **85–86**
abdominal thrusts, **143**
accidents, **269, 270**
acid indigestion, **189**
acids, stomach, **189–191, 206, 264**
acne, **22–23**
acquired immunity, **210–213**
activity levels, evolutionary and, **57–63**
adaptive immune system, **209–210**
adipose tissue, **40**
aerobic exercise, **72, 73, 77, 163, 172–178**
aerobic heart rate zone, **177, 178**
African countries, lifespan in, **259**
aging, **265–268**
    abnormally-fast, **266**
    bad advice for, **267–268**
    cancerous mutations and, **228**
    dealing with, **266–268**
    erectile dysfunction and, **240**
    eyesight and, **111**
    hearing and, **117**
    mechanisms of, **265–266**
    muscles and, **80**
    old age and death, **257**
    quality of life, **261**
    remaining life expectancy, **259–261**
    wrinkles, **19–21**
agonal phase of death, **273**
AIDS, **222, 225, 236**
air
    indoor air quality, **152–153**
    warming through nose, **141**
air cleaners and fresheners, **153**
air pollution, **150–156**
air-quality index, **151**

alcohol, **188, 240, 254**
allergens, **154**
allergies
    antibiotics and, **198**
    breastfeeding and, **248**
    colds and, **225**
    dust mites, **11, 12**
    environmental causes, **232–233**
    symptomology, **231–233**
aluminum, **30**
alveoli, **135, 151**
Alzheimer's disease, **269, 272**
ambulances, calling, **170**
American Association of Pediatrics, **236**
amino acids, **33, 192**
anaerobic exercise, **73**
anaphylaxis, **231**
angioplasty, **171**
anosmia, **126**
antacids, **189**
anti-aging products, **267**
antibacterial oil, **206**
antibacterial products
    allergies and, **232**
    soap, **214, 216**
antibiotics
    acne and, **23**
    allergies and, **232**
    antibiotic resistance, **218**
    bacteria and, **198, 218–219**
    withholding, **272**
antibodies, **209, 211, 248**
antihistamines, **231**
antioxidants, **21**
antiperspirants, **29–30**
anxiety. *See* stress
aorta, **167**
aphrodisiacs, **253**
apocrine sweat glands, **27, 28**
apoptosis, **226**
appetite
    fat levels and, **41**
    weight processes, **55–63**
apple-shaped bodies, **52**

arousal, 252–253
arteries, 155, 161, 165–178, 266
arthritis, 42, 98
Asian countries, lifespan in, 259
aspirin, 171
asthma
    antibiotics and, 198
    causes of, 154
    colds and, 225
    dust mites and, 11, 12
    sterile environment and, 232, 233
astigmatism, 109–110
athletes
    body-fat percentage, 48
    types of muscles, 68
atria, heart, 159
auricles, 118
autism, 211
autoimmune diseases, 230–231

B

B cells, 209
babies
    age of average death and, 262
    baby formula, 248
bachelors, 215
back. *See* spine
bacteria
    after death, 274
    antibiotics, 218–219
    bad breath and, 185
    balancing, 198–199
    benefits of, 214
    body odor and, 27, 29
    in breast milk, 248
    co-existing with, 212
    diarrhea, 197
    environmental niches, 215–217
    kissing and, 252
    large intestine, role in, 196–198
    macrophages and, 207
    in mouth, 141, 182
    mucus and, 144
    as pathogen, 206
    probiotics and prebiotics, 199
    raw cookie dough, 217
    size of, 220
    skin bacteria, 214
    spoiled food, 219
    stomach acids and, 189
    superbug, 218
    tips for cleaning, 216
    typical weight of, 213
bad breath, 141, 185
balance, inner ear and, 121

baldness, 36
barcode men, 36
bench presses, 84
benzoyl peroxide, 23
biceps, 69
bicycling, 174
bile, 193
biofeedback, 161
birth control, 249, 251
bitterness, 128
Blaine, David, 137
bleach, 216
blindness, 110
blind spots, 114–116
blisters, 19
blood. *See also* circulatory system
    amounts of, 160, 164
    blood types, 166
    color of, 164
    contents of, 164–165
    cooling through sweating, 26
    full circuit through body, 168
    heart pumping process, 159
    map of circulatory system, 167
    in stool, 229
    veins and arteries, 165–168
blood cells, 91, 95, 136
blood flow
    temperature control and, 24
blood pressure
    aging and, 268
    checking, 50
    cigarettes and, 155
    erectile dysfunction and, 240
    heart attacks and, 171
    lowering, 163
    measuring, 161–163
    sex and, 252
blood sugar, 50, 195
blood types, 166
blood vessels, 165–168
blowing one's nose, 145, 225
blue balls, 239
blushing, 25
BMI (body mass index), 46–47
body-fat percentage, 47–49
body-fat scales, 49
body mass index (BMI), 46–47
body odor, 27–30
bones
    aging and, 266
    bone scans, 94
    functions of, 91
    imagination and, 90
    joints and, 95–98
    as living tissue, 89

male and female differences, 91
marrow, 94–95
middle ear, 119
number of, 90
preventing bone loss, 93–94
rebuilding bones (remodeling), 92
spine, 98–104
borborygmus, 188
Botox injections, 20
bowels. *See* intestines
brain
    aging and, 266
    Missing Manual for, 4
    near-death experiences and, 274
    strokes, 169
breast milk, 211
breasts, 229, 247–249
breathing
    asthma, 154
    bad breath, 185
    choking hazards, 143–144
    cigarettes and, 155–156
    exercises, 138–139
    gas exchange in, 135–138
    holding one's breath, 137
    mucus and, 144–145
    nasal cycle, 142–143
    pollution and, 150–156
    process, 138–139
    respiratory system, 134–137, 140–145
    snoring, 140
    strength training and, 79
    throat and esophagus, 143–144
    voice characteristics, 145–149
Bristol stool chart, 200
broad-spectrum antibiotics, 218
bronchioles, 135
bruises, 19
brushing teeth, 183

C

caffeine, 94, 203
calcium, 13, 94
calories
    caloric restriction, 190, 265
    cravings for, 194
    fat and, 38, 39
    in semen, 242
campylobacter, 219
cancer. *See also* specific kinds of cancer
    as cause of death, 269
    immune system and, 226, 228–233
    older age and, 264
    prevention, 229
    stages of, 226–227
    young adult deaths, 270

capillaries, 167
carbohydrates, 188, 191, 192, 195, 248
carbon dioxide, 136, 137, 151, 165, 167
carbon monoxide, 137, 156
carcinogens, 155
cardiac arrests. *See* heart attacks
cardiac muscles, 67
cardiac stress tests, 175
cardio exercises, 172–178
cardiovascular system. *See* circulatory system;
            heart
carpal tunnel syndrome, 71
carpet, 153
cartilage, 96, 99
castration, 237
cataracts, 112
cavities, 182
cells, 9
    after death, 273
    aging and, 265
    B cells, 209
    blood cells, 91
    cancer and, 226–229
    germ-line cells, 223
    light-sensing cells, 112
    macrophages, 207
    marker molecules in, 230
    memory cells, 210–213
    natural killer cells, 228
    recognition, 230
    self-destruct sequence in, 226
    T cells, 209
    viruses and, 220–225
cellulite, 51
cellulose, 193
central nervous system, 230
cervical curve, 99
cheese, 192
chemical peels, 21
chemicals
    digestive, 180
    flavor and, 131
    smell and, 125
chest pain, 169
chewing food, 181
chicken meat muscle comparisons, 68
childhood
    body odor and sweat glands, 28
    controlling fat, 45
    developing new taste preferences, 129
    fat cell development, 44
    sun exposure during, 15
choking, 143, 146
cholera, 222
cholesterol, 50, 171, 268

cigarettes
  bone health and, **94**
  effects of, **155–156**
  heart attacks and, **172**
  indoor air pollution, **152**
  lung cancer and, **229**
  lung health after quitting, **156**
  wrinkles and, **20**
cilia, **151, 156**
circulatory system
  blood, **164–165, 168**
  blood pressure, **161–163**
  cardio exercises, **172–178**
  heart, **158–163**
  heart attacks, **168–172**
  map of, **167**
  veins and arteries, **165–168**
circumcision, **236**
cleaning products, **152**
cleanliness
  air quality after housecleaning, **153**
  allergies and, **232**
  bacteria and, **197, 214**
  dust mites, **12**
  food-borne illnesses and, **198**
  preventing bacterial problems, **216**
  shampoos and conditioners, **34**
  skin care, **8**
clitoris, **246–247, 254**
clotting, **165, 169**
cochlea, **121, 122**
coffee, **203**
colds
  average per year, **224**
  benefits of, **225**
  flavors and, **131**
  mucus and, **144**
  profile of, **223–225**
  symptoms, **224**
cold temperatures
  dust mites and, **12**
  hats and heat loss, **24**
  skin and, **23–30**
colitis, **230**
collagen, **18, 21**
colon, **195–204**
colon cancer, **228**
colon "cleansing", **190**
colonoscopies, **229**
color
  of blood, **164**
  color vision, **112–113, 114**
  of food, **132**
  of stool, **200**
  of semen, **242**
color-blindness, **113**

combinations of foods, **132**
comb-overs, **36**
common cold. *See* colds
complex carbohydrates, **195**
compound exercises, **77**
concentric contractions, **79**
conception, **254**
conditioners, **33–34**
condoms, **192**
cones, **112–113**
connective tissue, **230**
constipation, **200, 201, 203**
contact lenses, **112**
contracting muscles, **66, 79**
control
  at end of life, **271**
  weight training, **78**
cookie dough, **217**
cooking, **153**
cool down periods, **76**
corneas, **108, 109–110**
corpus cavernosa, **238**
cortisol, **63**
coughing, **229**
countertops, **215**
cracking knuckles, **97**
Crohn's disease, **230**
cross-country skiing, **174**
cultural factors
  changes in diets, **58**
  weight issues, **57–59**
cuts, **19**
cutting boards, **215, 216**
cycling, **174**

D

dairy products, **145**
dander, **152, 231**
DASH diet, **163**
death
  decay and, **274**
  dying with dignity, **270–272**
  life expectancy and, **258–261**
  most common causes, **269–270**
  mutations and, **264**
  naturalness of, **261–263**
  near-death experiences, **274**
  old age and, **257**
  reasons for, **262**
  remaining life expectancy, **259–261**
  sex and death theory, **264**
  stages of, **272–274**
death rattle, **273**
decaying process, **274**

decibel scale, 123

deep breathing, 138

dehydration, 203

Demodex, 32

dentin, 182

dentistry, 182

deodorants, 29–30

depression, 240

dermis, 9, 18–21

detoxifying, 190

deviated septum, 141

diabetes, 50, 185, 240, 269, 270

diaphragm, 138

diarrhea, 197, 200

diastolic pressure, 161, 162

diet

  blood pressure and, 163

  immune system and, 209

  indigestible foods, 193

  life expectancy and, 265

dieting

  changes in cultural diets, 58

  difficulties of, 45

  effective techniques, 62

  extreme chewing diet, 181

  fasting and detoxifying, 190

  forced starvation diet, 265

  pros and cons of, 60–63

  understanding weight processes, 55–63

digestive system

  autoimmune diseases and, 230

  enzymes, 185

  indigestible materials, 193

  large intestine, 195–204

  length of, 179

  meal digestion times, 200

  mouth, 180–181

  small intestine, 190–195

  stomach, 186–190

dignity, dying with, 270–272

DNA, 220, 222, 223, 265

doctors

  fat and weight consultations, 49–50

  medical interventions, 272

  physicals, 229

doorknobs, 215

double-jointed people, 97

Down's syndrome, 245

drinking water, 203

drugs

  aging and, 268

  direct absorption of, 188

  erectile dysfunction and, 240

  hair traces of, 31

dry-cleaned clothes, 153

Dukoral, 197

dust

  controlling, 12

  dusting and air quality, 153

  mucus and, 144

  particulate matter, 150

  skin cells as, 10

dust mites, 11–12, 220

dyeing hair, 35

dying. *See* death

E

ears

  aging and, 117

  air vibrations and, 115–116

  ear canal, 118

  ear hairs, 33

  eardrum, 119

  inner ear, 121–122

  interpreting sound, 116–117

  middle ear, 119–120

  outer ear, 118–119

  volume, 123–124

earwax, 118, 119

eating. *See also* diet; dieting

  aging and, 267

  best foods for digestion, 194–195

  fasting and detoxifying, 190

  mouth, chewing and teeth, 180–186

  psychology of, 63

  stomach, 186–190

  understanding weight processes, 55–63

Ebola virus, 224, 225

eccentric contractions, 79

eccrine sweat glands, 27

E. coli bacteria, 197, 212, 219

ecosystems (bacterial), 214

eggs (chicken), 217

eggs (human), 244, 250

ejaculation, 239, 242–243, 253

elastin, 18

electrical conductance fat measurements, 48

electron microscopes, 220

emergency rooms, 170

emphysema, 151

enamel, 181, 182, 184

endorphins, 274

endurance exercise zone, 177

energy

  converting sugar to, 134

  fat as storage system, 38

environment, allergies and, 232–233

enzymes
  pancreatic, **192**
  saliva and, **185**
  small intestine and, **192**
  stomach acids and, **189**
epidermis, **9, 13–17**
epiglottis, **143**
erections, **238–240, 255**
esophagus, **143**
estrogen, **244**
Eustachian tubes, **120**
evolution
  activity levels and, **57**
  bacteria and, **213**
  breasts and, **249**
  death and, **262, 263**
  exaptation, **145**
  food choices and, **194**
  hymen and, **246**
  near-work and eyesight, **111**
  penis size and, **241**
  sun's role in, **16**
  viruses and, **223**
exaptation, **145**
excrement
  blood in, **229**
  of dust mites, **12**
  fiber and, **200–204**
  passing stool, **199–200**
exercise
  aging and, **267**
  back exercises, **103–104**
  blood pressure and, **163**
  bone health and, **93**
  breathing exercises, **138–139**
  building muscle, **73–75**
  cardio exercises, **172–178**
  evolutionary changes, **57**
  heart rates during, **160, 174–178**
  immune system and, **209**
  muscle pairs and, **69**
  personal trainers, **87**
  schedules and routines, **76–77**
  six essential exercises, **80–88**
  stretching, **75–76**
  types of, **72–73**
  voice exercises, **147–149**
exfoliation, **8**
exhaustion, point of, **78**
eyebrows, **32, 35**
eyelashes, **32, 33, 35**
eyes
  anatomy, **108–109**
  bacteria in, **213**
  binocular vision, **113–115**
  blind spots, **114–116**
  color and night vision, **112–113**
  eye charts, **110**
  eye movement and scanning, **107**
  focusing problems, **109–112**
  vision, **107**

F

facial muscles, **20, 71**
fallopian tubes, **244, 251**
farsightedness, **109–110**
fasting, **190**
fast-twitch muscle fibers, **67–68, 73**
fat and fat cells, **37**
  body-fat percentage, **47–49**
  body mass index (BMI), **46–47**
  body shapes and, **52–53**
  in breast milk, **248**
  cholesterol tests, **50**
  consuming during diets, **61**
  cravings for, **129**
  creation of new cells, **42–45**
  cultural factors, **57–59**
  digesting, **188, 193**
  doctor's consultations, **49–50**
  fat cell functions, **38, 40–63**
  hormone production and, **40–42**
  indigestible, **193**
  life cycle of cells, **44**
  measuring, **45–50**
  moral issues, **45**
  myths about, **39**
  numbers of fat cells, **40, 42–63**
  pros and cons of diets, **60–63**
  radioactively dating cells, **42–63**
  role in hormone production, **40–42**
  small intestine processing, **191**
  taste of, **128**
  types of fat cells, **51–63**
  unsaturated fat, **188**
  weight processes and, **55–63**
fat-burning exercise zone, **177**
fatigue, **169**
fatty acids, **193**
feces. *See* excrement
female reproductive organs. *See also* sex
  anatomy, **244–246**
  breasts, **247–249**
  clitoris and pleasure, **246–247**
  fertility and ova production, **245**
  G-spot, **248**
  hymen, **246**
  menopause, **245**
  menstruation. *See* menstruation
  tubal ligation, **251**
  uterus, **249–250**
  vagina, **245**
fertility, **41, 250**

fevers, 208, 229
fiber, 188, 193, 200–204
fibrin, 165
fight-or-flight response, 25, 165
filtering air, 153
finger musculature, 70
fingerprints, 27
fitness zone, 177, 178
flatulence, 202
fleas, 12
Fletcher, Horace, 181
flossing teeth, 183
flu, 225, 269
fluoride, 184
focusing, anatomy of, 108
food. *See also* digestive system
    acne and, 23
    adding good foods to diet, 61
    aging and, 267
    appearance, 132
    best foods for digestion, 194–204
    calcium-rich, 94
    changes in cultural diets, 58
    combinations, 132
    eating spoiled food, 219
    environmental influences, 59
    expectations of, 132
    indigestible, 193
    life expectancy and, 258
    mouth, chewing and teeth, 180–186
    portions of, 189
    preventing bacterial infections, 216
    rancid, 132
    temperature of, 131
    texture of, 131
    varying foods, 194
    weight processes and, 55–63
food-borne illnesses, 198
foreplay, 248, 252–253
foreskin, 236
form, strength training, 78
formula, 248
free radicals, 265
French paradox, 58
Freud, Sigmund, 236
friction ridges, 27
frizziness of hair, 31
frostbite, 24
Fujita, Koichiro, 233
fungi, 218
fur, 26, 31

G

G-spot, 248
gallbladder, 191, 193

garlic, 185
gas, 202
gasping, 177
gastrointestinal system. *See* digestive system
gender. *See* men; women
genes
    aging and, 266
    autoimmune diseases and, 231
    viruses and, 220, 222, 223
    weight and, 57, 59
    wrinkles and, 20
genitalia. *See* reproductive system
genital retraction syndrome, 240
germ-line cells, 223
gingivitis, 184
glands. *See* names of specific glands
glasses, 112
glucose, 134
glycerol, 134
glycogen, 60
golden hour for heart attack, 168
goose bumps, 31
gorillas, 241, 243
The Great Masticator, 181
gum disease, 184
gyms, 87, 174

H

hair, 30–36
    after death, 273
    aging and, 266
    comb-overs, 36
    dyes, 268
    evolution of, 16
    frizziness, 31
    growth of, 34
    hair shafts, 30
    hair-transplant operations, 36
    keratin, 9
    loss of, 34–36
    removal, 35
    shampoos and conditioners, 33–34
hands
    bacteria on, 214
    washing, 216
hardening of the arteries, 155
Hashimoto's disease, 230
hay fever, 233
HDL cholesterol, 50
head, heat loss, 24
head voice, 147
"health gain" concept, 54
health gurus, 268
healthy-heart zone, 176, 177, 178
healthy life expectancy, 261

hearing
  age and, 117
  air vibrations and, 115–116
  inner ear, 121–122
  interpreting sound, 116–117
  middle ear, 119–120
  outer ear, 118–119
  volume, 123–124
hearing tests, 117
heart. *See also* circulatory system; heart attacks;
       heart rates
  aging and, 266
  amount of blood pumped, 160
  blood pressure, 161–163
  blood's full circuit through body, 168
  cardiac muscles, 67
  cardio exercises, 172–178
  causes of death and, 269
  flexibility and fragility, 158
  heart attacks, 168–172
  heartbeat and pumping, 159–160
  heart murmurs, 160
  heart rates, 160–161
  lifetime heartbeats, 158
  map of circulatory system, 167
  preventing attacks, 171–172
  scar tissue, 73
heart attacks
  aspirin, 171
  cigarettes and, 172
  grim statistics on, 168
  preventing, 171–172
  smoking and, 155, 156
  symptoms, 169–171
  triggered by other illnesses, 270
heart rates
  biofeedback, 161
  exercising and, 174–178
  measuring, 175
  recovery heart rate, 178
  variations and levels of, 160–161
heartbeats, 159–160
heartburn, 189
heat. *See* hot temperatures
Heimlich maneuver, 143
hemochromatosis, 264
hemoglobin, 136
hemorrhoids, 201
HEPA filters, 153
high blood glucose, 50
high blood pressure, 50, 162, 163, 240
histamine, 231
HIV, 222, 225
holding one's breath, 137
hormones
  acne and, 22
  in blood, 165

erectile dysfunction and, 240
estrogen, 244
fat cells and, 40–42, 42
in semen, 242
hospitals, 170
hot temperatures
  dust mites and, 12
  food and, 131
  skin and, 23–30
  sweating, 25–27
H. pylori, 207
hunger. *See* appetite
hydrochloric acid, 189
hydrostatic weighing, 48
hygiene. *See* cleanliness
hygiene hypothesis, 154
hymen, 246
hymenorraphy, 246
hypertension, 162, 163
hyperventilating, 137
hypodermis, 18
hypotension, 162

I

ice machines, 215
immune system
  acquired immunity, 210–213
  adaptive immune system, 209–210
  allergies and, 231–233
  antibiotics, 218–219
  autoimmune diseases, 230–231
  blood's role in, 165
  breast milk and, 211
  cancer and, 226–229
  co-existing with bacteria, 212–219
  common cold, 223–225
  fat levels and, 41
  inflammation as defense, 207–208
  older age and, 264
  pathogens and, 206
  physical defenses, 206
  sex and, 252
  strengthening, 209
  viruses and, 220–225
immunization, 211
immunotherapy, 231
inadequacy, 241, 249
indigestible foods, 201
infections, 154, 248, 270
inflammation
  aging and, 265, 270
  allergic, 231
  caused by fat, 41
  immune defenses, 207
influenza, 225, 269

inhalers, 154
inner ear, 121–122
insects, 11–36
insoluble fiber, 202
instrument strings, 192
insulin, 50, 188
intercourse. *See* sex
intestines
    animal, uses of, 192
    bacteria in, 213
    colon cancer, 228
    colon "cleansing", 190
    colonoscopies, 229
    intestinal gas, 202
    large intestine, 195–204
    small intestine, 190–191
intramuscular fat, 51
invasive treatments, 272
irises, 109
iron, 264

J

Japanese term for comb-overs, 36
jogging, 173
joints, 95–98, 230
jumping rope, 173
junk DNA, 223
junk food, 61

K

Kegel exercises, 255
keratin, 9, 30
kidneys, 185, 266, 269
kissing, 252
kitchen sinks, 216
knee musculature, 70
knuckle cracking, 97
Kosrae, 58

L

lactic acid, 76, 199
lactose, 199
lanolin, 22
large intestine
    bacteria's role in, 196–198
    balancing bacteria levels, 198–199
    digestive timing, 195
    intestinal gas, 202
    length of, 195–204
    passing stool, 199–200
laryngitis, 147
larynx, 146

laser resurfacing, 21
laxatives, 202
LDL cholesterol, 50
learned associations, 126, 129
left atrium and ventricle, 159
lenses, 108, 111
Leone, Mecco, 90
leptin, 41
lice, 12
licking wounds, 186
life
    animal lifespans, 262, 263
    dealing with aging, 266–268
    diets and lifespan, 190, 265
    healthy life expectancy, 261
    lifespans, 258–261
    maximum lifespan, 260
    mutations that shorten, 264
    remaining life expectancy, 259–261
    youth culture, 268
lifting heavy items, 102
lifting weights. *See* strength training
ligaments, 97
lighting, eyesight and, 112
light-sensing cells, 112
liposuction, 55
lips, 147
liver, 191, 193, 195, 264
liver disease, 269
looking young, 268
losing weight. *See* weight loss
loudness, 116, 123–124
low blood pressure, 162
lumbar curve, 99
lumps, 229
lung cancer, 151, 229
lung disease, 269
lunges, 82
lung packing, 137
lungs
    aging and, 266
    asthma, 154
    bad breath and, 185
    breathing exercises, 138–139
    breathing process, 138–139
    cigarettes and, 155–156
    gas exchange in, 135–138
    healing, 156
    holding one's breath, 137
    lung-function test, 154
    map of circulatory system, 167
    pollution and, 150–156
    respiratory system, 134–137
    total surface area of, 135
    upper respiratory system and, 140–145
    voice characteristics, 145–149

lupus, 230
lymph nodes, 168, 227, 229
lymphocytes, 209

M

macrophages, 207
macular degeneration, 112
magnetic resonance imaging (MRI), 48
male-pattern baldness, 36
male reproductive organs. *See also* sex
    anatomy, 236–238
    circumcision, 236
    ejaculation, 242–243
    erectile dysfunction, 240
    erections, 238–240
    orgasms and, 253–256
    penis, 236–238
    refractory periods, 254
    size and, 240–241
    testicles, 237
    vasectomies, 251
mammalian diving reflex, 137
marker molecules, 230
marrow, 94–104
masking sound, 122
mattresses, 12
maximum heart rate, 175
maximum lifespan, 260
measles, 211, 222
measuring fat, 45–50
meat, 194, 216
medical care, life expectancy and, 258
medical practitioners. *See* doctors
medications. *See* drugs
melanin, 16
melanoma, 14–36, 16–36
memory cells, 210–213
memory, odors and, 126
men
    average lifespan, 259
    bachelors and bacteria, 215
    body-fat percentage, 48
    heart rates, 160
    life expectancy, 259
    male-pattern baldness, 36
    male reproductive organs, 236–243
    sex and foreplay. *See* sex
    skeletal differences, 91
    vasectomies, 251
menopause, 245
menstruation, 28, 35, 245, 249, 250
mercury, 211
metabolism, 39, 55–63
metastasizing cancer, 227
microdermabrasion, 21

micropenis, 241
middle ear, 119–120
milk, 145, 189, 248
minerals, 92, 248
mirror exercise for voice, 148
mitochondria, 67, 75, 213
moisturizers, 8, 21
mold, 150, 152, 219
moles, 16–17
monogamy, 243
morbidly obese designation, 46
morning breath, 185
mouth
    bacteria in, 182, 213
    bad breath, 185
    breathing and, 141–143
    dental maintenance, 182–185
    flavors and, 131–132
    nighttime breathing, 141
    role in digestive system, 180–181
    saliva, 185–187
    supertasters, 130
    taste, 127–132
    teeth, anatomy of, 181–182
mouth-to-mouth resuscitation, 136
MRI (magnetic resonance imaging), 48
mucus
    common cold and, 225
    as defense system, 206
    filtration properties, 141
    middle ear, 120
    myths, 145
    normal daily amounts, 144
    stomach lining, 188, 190
    in upper respiratory system, 144–156
multiple sclerosis, 230
mumps, 211
murder, 269
muscles
    aging and, 80
    breathing and, 138
    building, 73–75
    chicken meat comparisons, 68
    creating muscle cells, 73
    defined, 66
    exercising, 72–80
    eye muscles, 107
    heart, 159, 169
    inner ear, 123
    loss of, 80
    muscle tone, 66
    pairing up actions, 68–69
    pelvic floor muscles, 255
    rectum, 199
    six essential exercises, 80–88
    skeletal muscles, 67–68

stomach, 188
strength training, 77–80
stretching, 75–76
tendons, 70–71
turning to fat, 39
types of, 66–67
mutations, 226, 228

N

nails, 9, 273
narrow-spectrum antibiotics, 218
nasal cavities. *See* nose and nasal cavities
nasal cycle, 142–143
natural foods, 194
natural killer cells, 228
naturalness of death, 261–263
nausea, 169
near-death experiences, 274
nearsightedness, 109–110
nicotine, 156
night vision, 112–113
nipples, 248
normal weight, 46
nose and nasal cavities
    anatomy of, 124–125
    blowing one's nose, 145, 225
    breathing and, 141–143
    flavors and, 131–132
    functions of smell, 126–127
    loss of sense of smell, 126
    mucus and, 144–145
    nasal cycle, 142–143
    nose hairs, 33
    odor fatigue, 127
    science of smell, 124
nostrils, 142–156
nuclear bombs, 42–63
Nurses' Health Study, 61

O

Obama, Barack, 148
obesity
    age-related fat cell development, 44
    as recent phenomena, 59
    BMI measurements, 46
    body-fat percentage, 47–49
    body mass index (BMI), 46–47
    body shapes and, 52
    cultural factors, 57–59
    doctor's consultations, 49–50
    measuring fat, 45–50
    moral issues, 45
    numbers of fat cells, 42

prevalence of, 37
pros and cons of diets, 60–63
research references, 57
obesity paradox, 61
odorant molecules, 125
odor fatigue, 127
odor receptors, 124, 125
odors, memory and, 126
office work, effects of, 101
oil glands, 21–23
old age
    aging, 265–268
    death caused by, 257, 270
    decline and death, 273–274
    looking old, 268
    quality of life, 261
oldest living person, 260
olestra, 193
olfactory bulb, 125
optical illusions, 115
optimism, 209, 267
orangutans, 243
O'Reilly Network Safari Bookshelf, 5
orgasms, 239, 246–256, 248, 253–256
ossicles, 119
osteoarthritis, 42, 98
osteoporosis, 93–104
outer ear, 118–119
ova, 244
ovaries, 244, 250
overexertion, 177
overhead presses, 83
overweight problems
    body-fat percentage, 47–49
    body mass index (BMI), 46–63
    body shapes and, 52
    cultural factors, 57–59
    doctor's consultations, 49–50
    measuring fat, 45–50
    moral issues, 45
    obese and overweight populations, 37
    pros and cons of diets, 60–63
    understanding weight processes, 55–63
ovulation, 244, 250
oxygen
    aerobic exercise, 73
    in blood, 165
    gas exchange in lungs, 135–138
    levels after quitting smoking, 156
    map of circulatory system, 167
    respiratory system, 134–137
ozone, 151, 153

P

pain
  eating and, 132
  heart attacks, 169, 170
  immune system responses and, 208
  ovulation, 250
  spine and posture, 99–103
  strength training and, 79
painkillers in breast milk, 248
paintings, cleaning with saliva, 186
pancreas, 191, 192
pancreatic cancer, 229
papillae, 127
parasites, 218, 233
Parkinson's disease, 269
particulate matter, 150, 150–156
passing stool, 199–200
pasteurized egg products, 217
pathogens, 206. *See also* bacteria; fungi;
        parasites; viruses
patience, weight loss and, 61
pear-shaped bodies, 52
pelvic tilts, 103–104
penile cancer, 236
penis, 236–238
  angle of, 239
  ejaculation, 242–243
  erectile dysfunction, 240
  erections, 238–240
  girth, 245
  vaginal size and, 245
perception, 105–132
performance-training zone, 177
periodontitis, 184
periods. *See* menstruation
peristalsis, 191
personal trainers, 87
perspiration. *See* sweat
pheromones, 28
philematologists, 252
physicians. *See* doctors
pillows, 12, 103, 189
Pima Indians, 58
pimples, 22–23
pitch, 116, 122
plant-based foods, 194
plaque (arterial), 169
plaque (dental), 182, 183
plasma, 164
platelets, 91, 165
pneumonia, 269, 270
point of exhaustion, 78
polio, 222
pollen, 150, 220
pollution, 150

posture, 99–104
prebiotics, 199
pregnancy
  blood types and, 166
  hair growth during, 35
  passing stool during labor, 198
  pelvic tilt exercise, 103
  preventing, 251
  weight and, 59
premastication, 252
presbyopia, 111
prisoner obesity studies, 56
probiotics, 199
progeria, 266
propylthiouracil, 130
prostate gland, 238, 242
protein
  dietary amounts of, 194
  digesting, 188, 192
  in breast milk, 248
  small intestine processing, 191
  stomach acids and, 189
  virus cells and, 221
psoriasis, 230
puberty
  acne, 22–23
  body odor and sweat glands, 28
  ovary's supply of eggs and, 245
pubic hair, 33
pulp, 182
pulse, 175
pupils, 109
pus, 208
push-ups, 86

Q

Quarles, Francis, 267

R

rabies, 222
racial characteristics, 16
radioactive carbon in fat cells, 42–45
rancid food, 132
raw cookie dough, 217
raw eggs, 217
reading, 107
reading glasses, 112
recovery heart rate, 178
rectum, 199
red blood cells, 91, 136, 220
red lights, 113
redline zone, 177, 178
red marrow, 95
reference material, 4

referred pain, 170
refractory periods, 254
remaining life expectancy, 259–261
remodeling bones, 92
rennet, 192
repetitive stress injuries, 71
replication of viruses, 222
reproductive system
    embarrassment about, 235
    female reproductive organs, 243–251
    male reproductive organs, 236–243
    starvation and, 265
research, scientific, 4
respiratory system
    anatomy and functions, 134–137
    asthma, 154
    breathing exercises, 138–139
    breathing process, 138
    cigarettes and, 155–156
    colds and, 223–233
    gas exchange in, 135–138
    holding one's breath, 137
    mucus and, 144–145
    pollution and, 150–156
    snoring, 140
    throat and esophagus, 143–144
    upper respiratory system, 140–145
    voice characteristics and, 145–149
resting heart rates, 160
retinas, 109
retroviruses, 223
rheumatoid arthritis, 230
rickets, 13
right atrium and ventricle, 159
rigor mortis, 273
ringtones, 117
RNA, viruses and, 220, 223
rods, 112–113
root canals, 182
rope skipping, 173
rubella, 211
running, 173

S

saccades, 107
Safari Books Online, 5
saliva
    functions of, 185–187
    normal amounts of, 185
    release during chewing, 180
salmonella, 217, 219
salt, 25, 26, 128, 129, 163
sarcopenia, 80
scar tissue, 73
schizophrenia, 230

scientific research, 4
scrotum, 238
sebaceous glands, 22–23
sebum, 22, 34
secondary sexual characteristics
    breasts, 248
    penis size, 241
    pubic hair, 33
sedatives, 240
seeing. *See* vision
self-respect, 268
semen, 242
semicircular canals (ear), 121
seminal vesicles, 242
sense organs
    aging and, 266
    hearing, 115–124
    number of senses, 106
    perception and, 105–107
    smell, 124–127
    taste, 127–132
    vision, 107–115
septic shock, 269
septum, 142
sets, exercise, 78
sex
    animal world unusual practices and, 255
    benefits of, 252
    breasts and, 247–249
    clitoris's role in, 246–247
    ejaculation, 242–243
    erectile dysfunction, 240
    erections, 238–240
    experimentation, 255
    female reproductive organs, 243–251
    foreplay, 252–253
    G-spot, 248
    hints for better sex, 254–256
    male reproductive organs, 236–243
    orgasms, 253–254
    refractory periods, 254
    sexual positions, 255
    size of penis, 240–241
sex-and-death theory, 264
shampoos, 33–34
shaving, 35, 36
shivering, 66
short-arm syndrome, 111
Sims, Ethan, 56
sinks, 216
sinuses, 225
sitting, effects of, 101
SIV virus, 222
sixth sense, 106
skeletal muscles, 66–69
skeleton. *See* bones

skiing, 174

skimming text, 107

skin
    aging and, 266
    bacteria on, 214
    blushing, 25
    body odor and, 27–28
    breathing in skin cells, 152
    caring for, 8
    cosmetics, 21
    dead cells, 8
    deodorants and antiperspirants, 29–30
    dermis, 18–21
    dust mites and, 11–12
    epidermis, 9–10
    examining, 10
    fingerprints, 27
    function of, 8–12
    glands in, 21–23
    goose bumps, 31
    hair and hair follicles, 30–36
    human vs. reptile, 10
    hypodermis, 18
    lifespan of cells, 9
    as physical defense, 206
    preventing sun damage, 14–15
    racial characteristics, 16
    shedding cells, 10
    sun and damage to, 13–17
    sunscreens, 15
    sweating, 25–27
    tanning, 16–17
    tattoos, 18
    temperature control, 23–30
    total square feet of, 7
    troubleshooting problems, 19
    vitamin D and, 13–14
    weight of, 7
    wrinkles, 19–21

skin cancer, 14–17, 16–36, 229

skin creams, 21, 22, 23

skinfold measurements, 48

sleep apnea, 140

sleeping
    breathing and, 141
    dieting and, 63
    good posture, 102
    immune system and, 209
    snoring, 140

slipped disks, 99

slow-twitch muscle fibers, 67–68, 73

small intestine
    best foods for, 194–195
    functions of, 191–193
    length of, 190

small penis syndrome, 240

smallpox, 210, 222

smell
    breathing and, 141
    descriptions of, 125
    flavors and, 131–132
    functions of, 126–127
    loss of sense of smell, 126
    nasal cycles and, 142–156
    nose and nasal cavities, 124–125
    odor fatigue, 127
    science of, 124
    semen, 242
    stool, 200

smog, 151

smoking. *See* cigarettes

smooth muscles, 66

Snellen charts, 110

snoring, 140

soap, 8, 214, 216

solitary old age, 267

soluble fiber, 202

sorites paradox, 36

sound. *See* hearing

sound waves, 116

sourdough, 199

sourness, 128

speaking. *See* talk test; voice

sperm, 242–256, 244, 250, 251

spices, 132

spinal cord, 98

spine
    back exercises, 103–104
    functions of, 98–99
    pain, 99–103
    posture, 99–103

spit. *See* saliva

Splenda, 193

spoilage bacteria, 219

sponges, 215, 216

squats, 81

stages of cancer, 227

staining, 30

standing posture, 100

starch. *See* carbohydrates

starvation
    extreme diets, 265
    studies, 56
    understanding weight processes, 55–63

static stretching, 75

step aerobics, 174

sterilizing kitchenware and sponges, 216

steroids, 154, 231

stomach, 187–191

stomach acid, 189, 206, 264

stool. *See* excrement

straight hair, 33

strength training
  defined, 72
  effective weight lifting, 77–80
  muscle loss and, 80
  scheduling, 77
  six essential exercises, 80–88
  women, 74
stress
  acne and, 23
  blood pressure and, 163
  erectile dysfunction and, 240
  immune system and, 209
  sex and, 252
  sweating and, 25
stretch marks, 19
stretch tolerance, 75
stretching, 75–88
strokes, 169, 269, 270
subcutaneous fat, 51
sucralose, 193
sugar
  converting to energy, 134
  cravings for, 129
  digesting, 195
  in blood, 165
  indigestible, 193
suicide, 269, 270
sun
  childhood overexposure, 15
  evolution of races and, 16
  preventing skin damage, 14–15
  skin cancer and, 229
  vitamin D and, 14
  wrinkles and, 20
sunscreens, 15
superbugs, 218
supertasters, 130
surfactants, 34
sweat
  amounts of, 26
  excessive, 30
  functions of, 25–27
  heart attacks, 169
sweat glands
  body odor and, 27–28
  functions of, 21–23
  preventing body odor, 29–30
  sweating, 25–27
  types of, 27
sweetness, 128, 129
swelling, 208
swimming, 173
synovial fluid, 97
systolic pressure, 161, 162

T

T cells, 209, 225
talking. *See* voice
talk test, 176, 177
tanning, 16–17
tanning beds, 15
tapeworms, 233
target heart rate, 176
tartar, 182
taste and taste buds, 127–132
taste tests, 130
tattoos, 18
tea, 203
teenagers. *See* puberty
teeth, 181–185
temperature
  skin, 23–36
  testicles, 238
tendinitis, 71
tendons, 70–88
terminal hair, 32
terminally-ill patients's desires, 270–272
testicles, 229, 237, 239, 242–256, 251
testicular cancer, 237
testosterone, 22, 74, 237, 240
texture of food, 131
thin people, 39, 54
third-hand smoke, 152
thirst, 203
thoracic curve, 99
threshold zone, 177, 178
throat, 140, 143–156, 213
thyroid, 230
tightness in chest, 169
tinnitus, 124
tiredness, cancer and, 229
tobacco. *See* cigarettes
tongue
  flavors and, 131–132
  speech and, 147
  supertasters, 130
  taste and, 127–132
toothbrushes, 215
toothpaste, 184
touch, 106, 132
toxins, 165, 190, 265, 267
trachea, 135
trans-fats, 61
traveler's diarrhea, 197
travelling tips, 12
treatable cancers, 227
triceps, 69
triclosan, 184, 216
triglycerides, 50
tubal ligation, 251

tumors, 226
twin studies, 56

U

ulcers, 189, 207, 264
ultraviolet rays, 13, 14, 34
umami, 128
underwater weighing, 48
underweight problems
  body-fat percentage, 47–49
  body mass index (BMI), 46–47
  measuring fat, 45–50
  understanding weight processes, 55–63
  visceral fat and, 54
unprocessed foods, 194
unsaturated fat, 188
untreatable cancers, 227
upper respiratory system, 140–145, 223
urethra, 238, 242
uterus, 244
UVA, UVB, and UVC rays, 14

V

vaccinations, 209, 210, 211
vacuuming, 153
vagina, 245, 248, 255
valves, 159, 166
varicose veins, 166
vas deferens, 251
vasectomies, 251
veins, 165–168
vellus hair, 32
vena cava, 167
ventilation, 152
ventricles, 159
vertebrae, 98
vertebral column, 98–104
Viagra, 240
videos, 5
virginity, 246
viruses
  anatomy of, 220
  antibiotics and, 218
  colds and, 223–225
  deadliest, 225
  immortality of, 221
  life cycles of, 221–223
  mucus and, 144
  retroviruses, 223
  size of, 220
  successful viruses, 224
  viral DNA, 223

visceral fat, 51, 52–55
vision
  binocular vision, 113–115
  blind spots, 114–116
  color and night vision, 112–113, 114
  defects in, 109–112
  eye anatomy, 108–109
  eye movement and scanning, 107
  eye muscles, 107
vitamin A, 209
vitamin B, 196, 209
vitamin B12, 187
vitamin C, 209, 224
vitamin D, 13–14, 94
vitamin E, 209
vitamin K, 196
vitamins
  anti-aging, 21
  hair and, 33
  in breast milk, 248
vitiligo, 230
vocal folds, 146–147
vocal nodules, 147
VOCs (volatile organic compounds), 152
voice
  exercises for, 147–149
  using breathing apparatus for, 145
  vocal folds, 146–147
volatile organic compounds (VOCs), 152
volume, 116, 123–124

W

waist measurements, 53
warming up, 75, 79
warts, 19
waste products. *See also* excrement
  aging and, 265
  in blood, 165
  detoxifying, 190
  map of circulatory system, 167
water
  absorbing, 200
  amount to drink daily, 203
  fat levels and, 49
  mixing with food, 195
websites
  Missing Manuals, 4
  O'Reilly Network Safari Bookshelf, 5
  reference material, 4
weight
  aging and, 267
  blood pressure and, 163
  body-fat percentage, 47–49
  cultural factors, 57–59

doctor's consultations, 49–50
of human body's bacteria, 213
measuring fat, 45–50
obese or overweight populations, 37
pros and cons of diets, 60–63
of skin, 7
weight loss
cancer and, 229
effective techniques, 62
fat cell numbers and, 44
"health gain" concept, 54
heart rates and exercise, 178
losing visceral fat, 53–55
pros and cons of dieting, 60–63
types of fat and, 52
understanding weight processes, 55–63
weight training. *See* strength training
Western diets, 58
white blood cells, 91, 207, 209
white coat syndrome, 163
whiteheads, 22
whitening teeth, 184
womb, 249
women
average lifespan, 259
bacteria counts and, 214
body-fat percentage, 48

female reproductive organs, 243–251
fertility, 41, 245
heart attacks and, 168
heart rates, 160
menopause, 245
menstruation, 28, 245, 249
osteoporosis, 93
scent of men's shirts and, 28
sex and foreplay. *See* sex
skeletal differences, 91
tubal ligation, 251
virginity, 246
weight training, 74
workouts. *See* exercise
World War II starvation studies, 56
wounds, 186, 208
wrinkles, 19–21

Y

yellow marrow, 95
youth culture, 268